SMALL GASOLINE ENGINES

GEORGE E. STEPHENSON

VNR VAN NOSTRAND REINHOLD COMPANY
NEW YORK CINCINNATI TORONTO LONDON MELBOURNE

Printed in the United States of America

Published in 1978 by Van Nostrand Reinhold Company
A division of Litton Educational Publishing, Inc.
135 West 50th Street, New York, NY 10020, U.S.A.

Van Nostrand Reinhold Limited
1410 Birchmount Road
Scarborough, Ontario M1P 2E7, Canada

Van Nostrand Reinhold Australia Pty. Ltd.
17 Queen Street
Mitcham, Victoria 3132, Australia

Van Nostrand Reinhold Company Limited
Molly Millars Lane
Wokingham, Berkshire, England

16 15 14 13 12 11 10 9 8 7 6 5 4 3 2 1

Library of Congress Cataloging in Publication Data

Stephenson, George E.
    Small gasoline engines.

    Includes index.
    1. Internal combustion engines, Spark ignition.
I. Title.
TJ790.S73 1978   621.43'4   77-29119
ISBN 0-442-27976-0

PREFACE

SMALL GASOLINE ENGINES is designed to provide instructional material on the construction, operation, care and maintenance, and application of small two- and four-cycle gasoline engines. It supplies basic instruction for those who intend to qualify as mechanics in this field, and for others who want a general understanding of this power source. The book is appropriate for automotive courses, industrial arts programs, adult education and extension courses, and special courses for mechanics of small engines.

Each of the twelve units treats a single, logical block of instruction. Review questions, discussion topics, and instructor demonstrations supplement the content coverage, giving direction to readers' work. Laboratory experiences, included at the end of most of the units, guide readers in the procedures of assembly and disassembly of small gasoline engines. The laboratory experiences are directly related to the content coverage of the unit; theory and practice are thus combined to provide a unified instructional program. The procedures given are fundamental and apply to all makes of small gasoline engines.

Certain changes have been incorporated in this revision of SMALL GASOLINE ENGINES to ensure a complete, up-to-date coverage of the subject. All of the content has been reviewed for technical accuracy, logical sequence of presentation, and readability. The following specific changes have been made in this edition:

- A new unit on safety has been added which includes safety attitudes, safety with gasoline, and safety with small gasoline engines.

- Many new illustrations and photos have been added to further readers' understanding of the written material. These include clearly labelled diagrams of carburetor operation and parts, a five-step illustration of the magneto cycle, a troubleshooting chart, and an expanded Appendix.

The author of this publication, George E. Stephenson, is a graduate of Stout State University (BS) and Colorado State University (MED). He has also authored *Power Technology* and *Drawing for Product Planning*. In 1967 he received the AIAA Outstanding Teacher of the Year Award — state of Illinois. For several years, the author has served as a supervising teacher for the Western Illinois University student teaching program. Mr. Stephenson is currently a power mechanics and general industrial arts teacher.

CONTENTS

UNIT 1 WORK, ENERGY, AND POWER

OBJECTIVES

After completing this unit the student will be able to:

- state and use the formulas for work, efficiency, power, and horsepower.
- state the difference between kinetic and potential energy, and demonstrate sources of each.
- discuss and demonstrate the concept of conservation of energy.
- show how losses due to friction can affect the efficiency of a machine.

Before starting a detailed study of the elements of small gasoline engines, it is necessary to review some of the terms which are basic to an understanding of mechanical power. What is power? What is energy? What is work? Are these terms all the same and if not how do they differ? It is the purpose of this unit to define and explain these terms, so that as they are used in later units their meaning will be clear.

WORK

Work is a scientific term as well as an everyday term. To many people, work means engaging in an occupation. One person may work hard in an office while another person may work hard carrying bricks on a construction job. Both individuals may come home tired but, scientifically, only the one carrying bricks has done much work.

Work, in the scientific sense, involves the moving of things by applying force. Work, then, is applying a force to cause motion or, in other words, motion caused by applying a force.

Measurement of Work

The common unit for measuring work is the foot-pound. The amount of work re-quired to lift one pound, one foot is expressed as one foot-pound. A person carrying 50 pounds of bricks up a 10-foot ladder does 500 foot-pounds of work. Work is, therefore, applying a force through a distance.

Work = Force x Distance
(Ft.-lbs.) = (Pound) x (Feet)

Figure 1-1 shows how to calculate the work involved in lifting a 20-pound weight 5 feet.

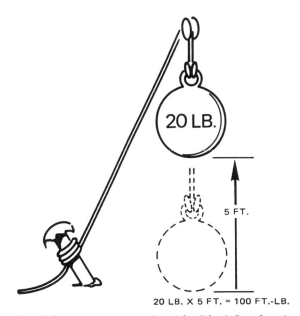

20 LB. X 5 FT. = 100 FT.-LB.

Fig. 1-1 A twenty-pound weight lifted five feet in the air

Work = Force x Distance
Work = 20 lbs. x 5 ft. = 100 ft.-lbs.

By scientific definition, the person who struggles to move a boulder but fails to budge it is not doing work because there is no "distance moved." Of course all engines can do work because they are capable of applying force to move machines, loads, implements, and so forth through distances.

ENERGY

Energy is the ability to do work. The energy stored in the human body, for instance, supplies the body with a potential to do work. A person can work for a long period of time on stored energy before needing to refuel with food. The human body is a good example of one major energy form: potential energy.

Potential Energy

Potential energy is the energy a body has due to its position, its condition, or its chemical state. The following are some examples of potential energy: water at the top of a waterfall (position); a tightly wound watch spring (condition); fuels such as gasoline, coal, etc., or foodstuffs to be burned by the body (chemical state).

Stored or potential energy is measured in the same way as work, in foot-pounds. How much potential energy is there when a 20-pound weight is lifted 5 feet?

Potential Energy = Force x Distance
Potential Energy = (weight) x (height)
Potential Energy = 20 lbs. x 5 ft.
Potential Energy = 100 ft.-lbs.

With the weight at the five-foot height, there is 100 foot-pounds of potential energy.

Kinetic Energy

Kinetic energy is the energy of motion. Some examples are the energy of a thrown ball, the energy of water falling over a dam, and the energy of a speeding automobile. Kinetic energy is, in effect, released potential energy.

Consider a thrown ball, the work of throwing it, and what becomes of it. If a 3/4-pound ball is thrown by applying a force of 8 pounds through 6 feet, how much work is done?

Work = Force x Distance
Work = 8 lbs. x 6 ft.
Work = 48 ft.-lbs.

What becomes of this energy? It exists as energy of motion of the ball, kinetic energy. The ball in flight has 48-ft.-lbs. of energy which it will deliver to whatever it hits. Refer back to the example of potential energy. When the 20-pound weight which rests 5 feet above the floor is allowed to fall, its potential energy of 100 ft.-lbs. becomes 100 ft.-lbs. of kinetic energy, figure 1-2.

Energy can change form but it cannot be destroyed. This is the *law of the conservation of energy*. Energy can take the form of

20 LB. X 5 FT. = 100 FT.-LB.

Fig. 1-2 A twenty-pound weight dropped to the floor from a height of five feet delivers 100 ft.-lbs. of kinetic energy.

light, sound, heat, motion, and electricity. Consider the potential chemical energy of gasoline in an engine. Upon ignition, its energy is converted mainly into heat energy. The heat energy is used to push the pistons down and is the kinetic energy of the rotating crankshaft.

Consider, again, the 20-pound weight at a 5-foot height. When the weight falls to the floor, its energy is not lost; it is transformed into other types of energy. The floor becomes slightly warmer at the point of impact due to the creation of heat energy. A sound can be heard at the moment of impact, indicating the presence of sound energy. Of course, the floor gives slightly, thus absorbing some energy.

The efficiency of machines that transform energy is not 100 percent. That is, some used energy does not perform a useful purpose. It may be wasted in moving the machine parts to overcome friction inherent in all machines. The *efficiency* of a machine is the ratio of the work done by the machine to the work put into it.

Stated as a formula:

$$\text{Efficiency} = \frac{\text{Output}}{\text{Input}} \times 100$$

or

$$\text{Efficiency} = \frac{\text{Input} - \text{Losses}}{\text{Input}} \times 100$$

or

$$\text{Efficiency} = \frac{\text{Output}}{\text{Output} + \text{Losses}} \times 100$$

The figure 100 in these formulas is a way of converting the fraction to a percentage.

POWER

In everyday language the word power can mean a variety of things: political power or financial power, for example. However, in the language of scientists and engineers, power refers to how fast work is done, or

how fast energy is transferred. *Power* is the rate of doing work, and the rate of energy conversion.

How fast a machine can work is an important consideration for engineers, and for consumers, too. For example, when gasoline is bought, it is a purchase of potential energy. The size and efficiency of the engine determine how fast this energy can be converted into power. A 10-horsepower engine can work only half as fast as a 20-horsepower engine.

Measurement of Power

Power is measured in foot-pounds per second or in foot-pounds per minute. Referring again to the 20-pound weight that is lifted 5 feet, how much power is required to lift the weight if it takes 5 seconds?

Work = Force x Distance

Work = 20 lbs. x 5 ft. = 100 ft.-lbs.

$$\text{Power} = \frac{\text{Work}}{\text{Time}}$$

$$\text{Power} = \frac{100 \text{ ft.-lbs.}}{5 \text{ seconds}}$$

Power = 20 ft.-lbs./second

As another example, if an elevator lifts 3500 pounds a distance of 40 feet and it takes 25 seconds to do it, what is the rate of doing work?

Work = Force x Distance

Work = 3500 lbs. x 40 ft. = 140,000 ft.-lbs.

$$\text{Power} = \frac{\text{Work}}{\text{Time}}$$

$$\text{Power} = \frac{140,000 \text{ ft.-lbs.}}{25 \text{ sec.}} = 5600 \text{ ft.-lbs./sec.}$$

Power can also be expressed in foot-pounds per minute. If a pump needs 10 minutes to lift 5000 pounds of water 60 feet, it is doing 300,000 ft.-lbs. of work in 10 minutes, which is a rate of 30,000 foot-pounds per minute.

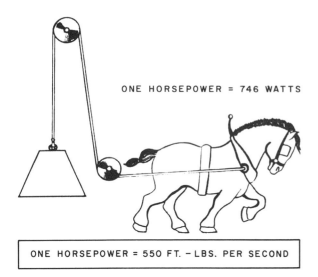

ONE HORSEPOWER = 746 WATTS

ONE HORSEPOWER = 550 FT. – LBS. PER SECOND

Fig. 1-3 Horsepower

Horsepower

The power of most machinery is measured in horsepower. The unit originated many years ago when James Watt, attempting to sell his new steam engines, had to rate the engines in comparison with the horses they were to replace. He found that an average horse, working at a steady rate, could do 550 foot-pounds of work per second, as shown in figure 1-3. This rate is the definition of *one horsepower.*

The formula for horsepower is:

$$\text{Horsepower} = \frac{\text{Work}}{\text{Time (in seconds)} \times 550}$$

If time in this formula is expressed in minutes, it is multiplied by 550 x 60 (seconds)

or 33,000. The formula, then, may be expressed as:

$$\text{Horsepower} = \frac{\text{Work}}{\text{Time (in minutes)} \times 33,000}$$

What horsepower motor would the elevator previously referred to have?

$$\text{Horsepower} = \frac{\text{Work}}{\text{Time (in seconds)} \times 500}$$

$$\text{Horsepower} = \frac{140,000 \text{ ft.-lbs.}}{25 \text{ sec.} \times 550}$$

$$\text{Horsepower} = 10 +$$

SUMMARY

This unit explained the terms which are basic to an understanding of the application of mechanical power that follow in later units. The measurements which these terms involve have been shown as formulas. To review, the following formulas are a means of expressing each of the terms covered:

WORK = FORCE x DISTANCE

$$\text{EFFICIENCY} = \frac{\text{OUTPUT}}{\text{INPUT}} \times 100$$

$$\text{POWER} = \frac{\text{WORK}}{\text{TIME}} = \frac{\text{FORCE x DISTANCE}}{\text{TIME}}$$

$$\text{HORSEPOWER} = \frac{\text{WORK}}{\text{TIME (in sec.)} \times 550}$$

$$= \frac{\text{WORK}}{\text{TIME (in min.)} \times 33,000}$$

REVIEW QUESTIONS

1. Give a definition of work.

2. What is the unit of measurement for work?

3. A 100-pound boy climbs a 16-foot stairway. How much work has he done?

4. Define energy.

5. Explain the difference between kinetic and potential energy.

6. Explain the conservation of energy. What are some common forms of energy?

7. Explain why a machine cannot be 100 percent efficient.

8. Define power.

9. What is the unit of measurement for power?

10. How much power is needed to lift a 100-pound bag of cement onto a 3-foot truck bed in 2 seconds?

11. What is the formula for horsepower?

12. How much horsepower does an engine on a grain elevator deliver if it can load 440 pounds of corn into a 25-foot storage bin in 10 seconds?

CLASS DISCUSSION TOPICS

- Have a student lift a given weight a definite height and calculate the work.
- Discuss the difference between scientific work and other kinds of work.
- Discuss and list many potential and kinetic energy sources.
- Discuss the origin of the term horsepower.

CLASS DEMONSTRATION TOPICS

- Demonstrate how the horsepower developed by a student can be calculated.
- Demonstrate the transformation of energy by striking and burning a match.
- Demonstrate potential and kinetic energy sources.
- Demonstrate the conservation of energy.
- Demonstrate losses due to friction and how they affect the efficiency of a machine.

UNIT 2 SAFETY

OBJECTIVES

After completing this unit the student will be able to:

- explain how one's personal attitudes toward safety affect other people.
- discuss the safe use and storage of gasoline.
- state methods of preventing carbon monoxide poisoning.
- summarize possible hazards in using small gasoline engines.

The maintenance, operation, repair, and testing which are involved in a study of small gasoline engines requires that careful and constant emphasis be placed on safety. It is beyond the scope of this unit to deal with the particular safety considerations of each different engine and its vast number of applications. Instead, this unit deals with safety considerations in several common areas: safety attitudes, safety with gasoline, carbon monoxide safety, safety with basic hand tools, safety with basic machines and appliances, and safety with small gasoline engines.

Specific safety information is included in later units as it applies, but it is useful to review this unit before beginning laboratory experiences.

SAFETY ATTITUDES

Safety is important to every student, both as an individual and as a member of a community. With the beginning of the machine age came many safety hazards. In transportation, alone, millions of power-driven vehicles presented a number of hazards. Industrial, construction, and scientific employees work with the tools of power technology every day. There are many hazards in the home, too—power appliances, household machines, fire, boiling liquids,

electricity, and so forth. Considering these hazards, it is necessary that an emphasis on safety begin early in childhood and continue throughout life.

Accidents can be compiled and shown in statistical form. Accidents can be measured in terms of frequency, probability, medical costs, time lost from work, or in many other ways that serve a useful purpose in identifying hazards and in developing safety programs.

On an individual basis, the statistical approach loses all of its importance. How can a number or a dollar sign be put on the personal cost and human suffering that result when accidents occur? People do not like to think of the unpleasantness of an accident happening to them. Fortunately, most try to protect themselves and those close to them from hazardous situations and accidents.

In gasoline engine safety, people must consider both their own safety, and the safety of others. Each person's actions and safety attitudes do have an effect on other individuals. Sometimes the effect is slow and indirect, such as parents' carelessness becoming part of their children's safety standards. At other times, the effect on others is direct, such as a head-on collision on the highway caused by an irresponsible driver. The point is that in power safety, individuals are responsible for themselves as well as for the safety of others.

Students of small gasoline engines should consider the safety aspect of every activity in which they engage. Most basic safety approaches to prime movers and machines are similar and the knowledge gained from one situation can be applied to another. Safety is never sacrificed for the sake of speed or expediency. It should have priority over everything else. How one operates a machine can reveal an individual's entire safety code, and indicate personal qualities or deficiencies.

Learning safety by trial and error alone is not enough. The safety considerations of each situation must be studied. A defensive attitude of, "if this happens, what might the result be?" must be a part of a person's overall judgment. Some safety is common sense, but many safety requirements must be learned from machinery instruction booklets and repair manuals. To ensure safety for oneself and others, one must know specific information that applies to the particular prime mover and its uses.

SAFETY WITH GASOLINE

Understanding the nature of gasoline and its safe use is very important. This fluid is used as a common item in many households. The presence of this liquid, which is more powerful than TNT, is readily accepted. Many people have become too casual about this volatile and explosive liquid.

Gasoline is around the home today (in the garage, toolshed, or barn) largely because the use of small gasoline engines has become so widespread. Lawn mowers, garden tractors, snow throwers, go-carts, utility vehicles, motorbikes, outboard motors, and chain saws need gasoline for operation and, as a result, a gasoline can may be found in nearly any home.

Accidents do happen; it is estimated that every year several hundred persons lose their lives due to accidents involving gasoline and other flammable liquids. In addition, several million dollars of property damage is caused by such accidents.

The volatility of gasoline is the characteristic that makes it such an ideal fuel but this same characteristic is also the prime danger. Gasoline can vaporize and explode in the atmosphere just as it can in the engine. A concentration of one to seven cubic feet of gasoline vapor per 100 cubic feet of air represents a flammable condition. Any spark or flame can touch off gasoline as a fire or explosion.

Gasoline should not be stored inside the home—it is too dangerous a liquid. It should be kept in a garage, toolhouse, or some other outside building. Storage in the basement is dangerous; a gas can could be tipped over and if the seal were imperfect, gasoline would leak out and begin to vaporize. Given certain circumstances, the whole basement could blow up—triggered by the flame from a furnace or water heater or from the arc of an electric switch.

Many states have recognized the hazards of gasoline around the home and are developing safety regulations that help to protect the public. The proper use of storage containers is one important area. Ideal storage is in a red metal can clearly labeled GASOLINE. Red is a universal color indicating danger. A metal can, such as shown in figure 2-1, will not

Fig. 2-1 Store gasoline and other flammable liquids in their proper containers.

not break if it is accidentally dropped or struck. Follow these rules for the storage of gasoline:

- Store gasoline in a metal container.
- Store gasoline in an outbuilding—not inside the home.
- Do not store gasoline during the off season. It is too dangerous to have around.
- Do not allow small children to have access to the gasoline supply.
- Store gasoline away from flames, excessive heat, sparks caused by static electricity, sparks caused by electrical contacts, or sparks caused by mechanical contact.
- Have the proper portable fire extinguisher available for use if needed.

Using gasoline safely involves decisions and judgments. It should be regarded as dangerous. Furthermore, it should be regarded as fuel for gasoline engines and not as a handy all-purpose solvent, cleaner, or fire starter. Again, the volatility and explosive nature of gasoline makes it unsuitable for any type of general household use. Gasoline used as a

Fig. 2-2 Have fire extinguisher readily available and know the correct use of the various types.

cleaning agent for paint brushes or to cut grease sets up real safety hazards. Accidental ignition can occur while the gasoline is being used, and after it has been dirtied, the problem of disposing of the liquid may be hazardous.

Never use gasoline to start a fire. The result may be a fire or explosion much larger than the intended size injuring the person with the gasoline.

Observe these precautions when using gasoline:

- Use gasoline only as a fuel for engines or devices where its use is clearly intended.
- Always regard gasoline as dangerous.
- Never smoke cigarettes, pipes, or cigars around gasoline.
- Never fill the tank of a hot engine or a running engine if the tank and engine are at all close to each other. Gasoline splashed on a hot engine can ignite or explode.
- When pouring gasoline from a container to the tank, reduce the possibility of a static electricity spark by having metal-to-metal contact. A metal spout held against the tank opening is good to use. A funnel that contacts both the container and the tank is also safe. Pouring gasoline through a chamois may present a static electricity hazard.

An understanding of fires and fire extinguishers, figure 2-2, is very useful information. Most fires fall into one of these categories:

Class A — Wood, cloth, paper, rubbish

Class B — Oil, gasoline, grease, paint

Class C — Electrical equipment

Class A Fires

Class A fires are the most common. Extinguishing these fires consists of quenching the burning material and reducing the temperature below that of combustion. Some extinguishing methods also have a smothering effect on the fire.

Most people are familiar with the soda acid extinguisher that is seen in many public buildings, industries, and places of business. To use this extinguisher it is simply upended; the water stream that is expelled should be directed at the base of the fire first, then back and forth following the flames upward. These extinguishers are not suitable for Class B fires such as gasoline fires. Foam extinguishers produce a foaming liquid when the extinguisher is inverted. For Class A fires, the foam should be directed at the base of the flames.

Class B Fires

The control and extinguishment of Class B fires, which include gasoline, presents a far greater hazard than that of Class A fires. Careless or unknowing action against this fire can place the fire fighter in great danger. Basically, the technique is to cut off the oxygen supply which feeds the burning liquid or to interrupt the flame.

Carbon dioxide (CO_2) extinguishers contain this gas under high pressure. Carbon dioxide does not support combustion and therefore has the effect of cutting off the oxygen in the air and smothering the flame. The extinguisher should be used with a slow sweeping action that travels from side to side, working to the back of the flame area. The discharge horn of the extinguisher becomes very cold during discharge and it should not be touched. Another hazard is that in a small room the extinguisher may produce an oxygen-short atmosphere that is dangerous to the fire fighter.

Foam fire extinguishers produce a water-base foam that can smother the fire. In using an extinguisher of this type on Class B fires, direct the foam at the back of the fire allowing the foam to spread onto the flame area. This minimizes the possibility of splashing the flaming liquid of an open container fire.

Class C Fires

Dry chemical extinguishers are excellent for Class B fires and for Class C fires. When the dry chemical is expelled under pressure, it interrupts the chemical flame chain reaction and thereby extinguishes the fire.

Class C fires involve electrical equipment and therefore water-base extinguishing materials are not suitable. Water on or around an electrical fire creates a shock hazard. If the electrical energy can be completely and positively shut off, the burning material can be handled according to its nature. If the equipment is energized, the fire must be fought with CO_2 extinguishers, dry chemical extinguishers, or vaporizing liquid extinguishers.

CARBON MONOXIDE SAFETY

The hazard of carbon monoxide is very real. Carbon monodide (CO) is a poison that is colorless, odorless and tasteless. It is a killer that is responsible for more poisoning deaths than any other deadly poison.

This gas is the result of the incomplete combustion of solid, liquid, or gaseous fuels of a carbonaceous nature. At home the gas is found along with improperly adjusted hot water heaters. In industry carbon monoxide can be found with kilns, oven stoves, foundries, smelters, mines, forges, and in the distillation of coal and wood, to list a few. However, it also exists in a much more common circumstance—the exhaust gases of internal combustion engines such as those used on the automobile. When an engine is in operation, carbon monoxide is produced. Knowledge about this poison and how to eliminate or minimize the danger is an important topic for everyone.

The action of this poison on the human body can be quite rapid. The hemoglobin of the blood has a great attraction for CO, three hundred times greater than that for oxygen. When the CO is combined with the hemoglobin it has the effect of reducing the amount of hemoglobin available to carry oxygen to the body tissues. If large amounts of CO combine with the hemoglobin, the body becomes starved for oxygen and literally suffocates.

Ventilation is of prime importance in preventing carbon monoxide poisoning. Unburned gases and exhausts must be carried away as effectively as possible by using chimneys, ventilation systems, and exhaust systems. These systems must be efficient because even small amounts of the gas can cause a dulling of the senses, which indicates danger.

Exhaust gases and carbon monoxide can get inside an automobile in several ways: through a defective exhaust system—tail pipe, muffler, or manifold; or through rusted-out or defective floor panels. If there is excessive CO around the automobile due to faulty exhausting or a poorly tuned engine, it can come in through an open window. For example, an open tailgate window on a station wagon can produce a circulation pattern that brings exhaust into the auto through the tailgate window. In the passenger section the carbon monoxide can build up to dangerous proportions. Station wagon rear windows should be closed, especially if other windows are open.

Consider the front window vents of the auto. When these are opened only slightly they draw air from the passenger compartment and if there is a hole at some exhaust concentration outside the compartment, exhaust can be drawn in. Generally, it is best to have the front vent windows open plus the side windows rolled down a bit in order to provide an adequate flow of fresh air. The air ducts that bring air in through the heater system are also a good ventilation source.

Knowing the symptoms of carbon monoxide poisoning and heeding their warnings could be a life-and-death matter. These symptoms include a tightness across the forehead, followed by throbbing temples, weariness, weakness, headache, dizziness, nausea, decrease in muscle control, increased pulse rate, and increased rate of respiration. If anyone has these symptoms while riding in a car, the car should be stopped, and the person should get some fresh air.

If a person is discovered who has passed out from carbon monoxide poisoning, remove the victim to the fresh air immediately. If breathing has stopped or if the victim is only gasping occasionally, begin artificial respiration at once. Also, have someone call a doctor and/or fire or emergency squad.

To prevent carbon monoxide poisoning, follow these rules:

- Do not drive an auto with all the windows closed.
- Do not operate internal combustion engines in closed spaces such as garages (figure 2-3, page 12) or small rooms, unless the room is equipped with an exhaust system.
- Keep carbon monoxide producing engines and devices tuned or adjusted properly in order to reduce the output of CO gas.
- Do not sit in parked cars with the engine running.
- Remember the importance of properly working ventilation and exhaust systems.

Fig. 2-3 **Do not underestimate the danger of carbon monoxide.**

SAFETY WITH BASIC HAND TOOLS

Tools should be treated with respect and should be well cared for. They should be stored in a way that will protect their vital surfaces, cutting edges, and true surfaces. A wall tool panel with individual hangers or holding devices for each tool is ideal, since the tools are easily seen, ready for use, and well protected. Tools carelessly thrown in a drawer can scratch, dull, or damage each other and are a hazard to the person who is looking for the right tool.

When tools must be carried in a toolbox, they should have individual compartments, if possible. If the box lacks individual compartments, the more sensitive tools and those with cutting edges should be wrapped in cloth.

Moisture and the rusting that results is a problem. If tools get wet, wipe them dry before storing. Also, the possibility of rusting can be lessened by wiping the tools with a slightly oily rag. If a tool picks up dirt during use, wipe it clean before putting it away.

The basic rules for the safe use of hand tools are

- Use tools that are in good condition and well sharpened. The dull tool is inefficient and is more likely to slip during use.

- Use the correct tool for the job and use it as it should be used. The incorrect use of tools causes many accidents.

In some operations, hand tools can present the hazard of flying metal chips. A mis-struck nail may ricochet across the room. Even a hammer can be dangerous. Especially dangerous are chipping operations with a cold chisel. All struck tools such as chisels and punches can mushroom out at the end over a period of use. See figure 2-4. These ends should be dressed on a grinder to eliminate the possibility of tool chips breaking and flying from the tool when it is hit. Hammers, themselves, can chip if they are struck side or glancing blows—hammers should strike the work with their full face, as shown in figure 2-5.

BEFORE DRESSING   AFTER DRESSING

Fig. 2-4 **All "struck" tools should be properly dressed on the end.**

Fig. 2-5 **The hammer face should strike the work with its full, flat face.**

When properly used, a wrench is a safe tool. But if used incorrectly, the wrench can be dangerous. Be certain to use the exact size wrench for the nut or bolt to prevent slipping or damage to the flats of the nut or bolt. If an adjustable wrench is used, snug it up tight against the flats of the nut or bolt and be sure that the direction of force places the major strain on the fixed jaw and not on the movable

WRONG

RIGHT

Fig. 2-6 Use an adjustable wrench so the strain is on the fixed jaw.

TOO SMALL

CORRECT SIZE

Fig. 2-7 Use the correct size screwdriver for the job.

jaw, figure 2-6. Most of all, pull a wrench toward you, do not push it. Pushing can be dangerous; the nut or bolt may suddenly loosen, or the tool may slip and the user be injured.

Screwdrivers are probably the most misused of all common tools; the list of abuses ranges from using it to open paint cans to using it as a wrenching bar. The tool is intended to be used in tightening or loosening various types of screws and therefore its tip must be preserved for this purpose. Use a screwdriver that is the correct size, figure 2-7.

It should fit the screw slot snugly in width and the blade should be as wide as the slot is long. Have several sizes of screwdrivers in a tool assortment. It is a poor practice to hold the work in one hand in such a manner that a slip would send the screwdriver into the palm of the hand. Screwdrivers, pliers, or other

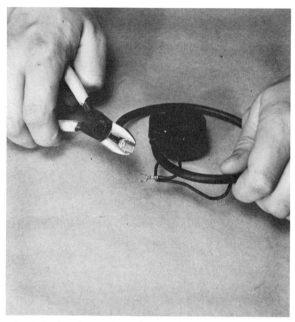

Fig. 2-8 Tools used in electrical and electronic work should have insulated handles.

tools that are used for electrical or electronic work should have insulated handles, figure 2-8.

Files are rarely sold with handles attached but they should be so equipped. The tang is fairly sharp and can cut or puncture the hand if the file suddenly hits an obstruction. The brittle nature of files makes them a poor risk and dangerous as a pry bar; do not use them for this purpose.

SAFETY WITH BASIC MACHINES AND APPLIANCES

Safe use of basic household machines, mechanisms, and appliances during normal use, adjustment, and repair is important to all. Reference is made to power hand tools; power tools; appliances such as stoves, refrigerators, dishwashers, clothes driers, washing machines, air conditioners, electronic equipment such as television sets, record players, and radios; powered hobby or sports equipment; small home appliances for kitchen or personal use; and so forth. Several common rules should be followed when using these machines:

- Before attempting to adjust or repair a machine, always unplug or disconnect the machine or device from the electrical power source. Be certain that someone else does not accidentally reconnect or plug in the machine while you are working on it. If the power source is in a remote location, tag the source "Do not connect. Equipment under repair."

- Before attempting to adjust or repair machines driven by engines, stop the engine, disengage the clutch, and remove the spark plug lead to eliminate the possibility of accidentally restarting the engine.

- Be sure to ground electrical equipment that requires grounding. Use the three-prong plug on portable tools or devices and properly ground the convenience outlet.

- Keep loose clothing away from rotating parts that could grab the cloth and thereby pull the operator into the machine.

- Protect eyes with safety glasses if there is any chance of flying chips or breakage.

- Thoroughly guard all belts, pulleys, chains, gears, etc. Do not remove these guards or allow them to be removed by others.

- Study the instructions that come with the machine. Be certain that the operating principles and the safety precautions are understood before operating or repairing the machine.

- Use the machine only in the manner and for the purpose for which it was designed.

- During adjustment and repair, re-tighten all nuts, bolts, and screws securely.
- Unless qualified to be there, keep away from high voltage areas in electronics equipment.

SAFETY WITH SMALL GASOLINE ENGINES

Safety education about lawn mowers, chain saws, sports vehicles, outboard motors, and other small gas engines should begin at an early age. These machines are very powerful and if improperly used can easily injure a person.

Small engine applications are often a person's first exposure to gasoline engines and powered machinery. There is danger in the fuel and its use around the machine, and there is danger in the use of the machine itself. Besides learning how to start and stop an engine, an operator must also know general safety considerations. Trial and error is a very poor way to learn the use of any prime mover or machine.

Anyone who operates a small gasoline engine should understand and use these safety rules:

- Do not make any adjustments or re-pairs to machinery being driven by an engine without first stopping the engine and removing the high tension lead from the spark plug. It is possible to accidentally turn the engine over and restart it, especially if the engine is warm. With the spark plug cable removed from the spark plug, the engine cannot start.
- Do not fill the gas tank when the engine is running or hot. If gasoline is spilled on a hot engine, a fire or explosion can result. If the job is not finished, and the engine needs more gasoline, stop the engine and let it cool off before refilling the tank. It is a good idea to start each job with a full gas tank.
- Do not operate a gasoline engine in a closed building, due to the carbon monoxide hazard.
- Read the equipment instruction booklet. Know and understand the equipment before operating it.
- Keep the equipment in perfect operating condition with all guards in place.

Power Lawn Mower

The power lawn mower is an example of one small engine application which is widespread in its use. Because of the widespread use certain safety procedures should be followed when operating the mower.

The cutting blade on lawn mowers can be dangerous. The whirling blade of a rotary lawn mower can cut through large sticks and pick up and hurl foreign objects like projectiles. Studies show that about half of the accidents that occur with rotary power lawn mowers are caused by thrown objects, and most accidents that involve the operator result in injury to toes, fingers, and legs.

Persons who operate a power lawn mower accept a safety responsibility for themselves and for others. These safety rules should be followed:

- When starting a lawn mower, be certain to stand clear with feet and hands away from the blade. See figure 2-9, page 16.
- Do not mow wet grass—it can be slippery. In the case of electric lawn mowers, wet grass can present an electrical shock hazard.

- Keep away from the grass discharge chute; sticks and stones are likely to be thrown out of this opening.
- When mowing hills or inclines, do not mow up and down; mow across the face of the slope.
- Do not pull a lawn mower—push it. If a person falls while pulling the mower, it may be pulled over the body.
- Inspect the lawn for rocks, sticks, wire, and other foreign objects that can be converted into projectiles by the mower blade.
- Allow only a responsible person to operate a power mower. Small children may be attracted to this type of activity, but should be kept away from the mowing area.
- Do not use a riding lawn mower as a play vehicle.
- Keep self-propelled lawn mowers under full control.
- Never leave a mower running unattended.

Fig. 2-9 Keep feet away from the blade when starting the mower.

REVIEW QUESTIONS

1. Explain why common sense precautions alone are not enough for the safe operation of an engine.

2. What characteristic of gasoline makes it such a dangerous liquid?

3. List several requirements for gasoline storage and precautions for gasoline use.

4. What type of extinguishers are best for putting out gasoline fires?

5. How is carbon monoxide produced?   What is its effect on the body?

6. Explain why it is important to disconnect electrical devices from their electrical source before repairs or adjustments are made.

7. Summarize the hazards in using hand tools.

8. List several safety rules for operating power lawn mowers.

CLASS DISCUSSION TOPICS

- Discuss how safety becomes a part of a person's personality.
- Discuss how every individual's safety attitudes can affect other persons.
- Discuss why learning safety through trial and error is a poor practice.
- Discuss the proper use and storage of gasoline.
- Discuss the various ways to prevent carbon monoxide poisoning.
- Discuss equipment instruction booklets and their importance to safe machine operation.
- Discuss the dangers of loose clothing to mechanics, technicians, or machine operators.
- Discuss the proper care and storage of hand tools.

UNIT 3 CONSTRUCTION OF THE RECIPROCATING ENGINE

OBJECTIVES

After completing this unit the student will be able to:

- identify the basic parts of an engine and explain how they work together.
- discuss how the valve train and camshaft work together, and identify the parts of each.
- explain the construction of the exhaust ports and show where they are located.
- identify the reed valves, and discuss their operation.

The internal combustion engine is classified as a *heat* engine; its power is produced by burning a fuel. The power stored in the fuel is released when it is burned. The word *internal* means that the fuel is burned inside the engine itself. The most common fuel is gasoline. If gasoline is to burn inside the engine, there must be oxygen present to support the combustion. Therefore, the fuel needs to be a mixture of gasoline and air. When ignited, a fuel mixture of gasoline and air burns rapidly; it almost explodes. The engine is designed to harness this energy.

The engine block is fitted with a cylindrical shaft. This is the *cylinder*. A plate fits over the head of the block, sealing off the top of the cylinder, the *cylinder head*. The cylinder contains a *piston,* a cylindrical fitting which fits the cylinder exactly. The piston is free to slide up and down within the cylinder. The fuel mixture is brought into the cylinder, then the piston moves up and compresses the fuel into a small space called the *combustion chamber.* When the fuel is ignited and burns, tremendous pressure builds up. This pressure forces the piston back down the cylinder; thus the untamed energy of combustion is harnessed to become useful mechanical energy. The basic motion within the engine is that of the piston sliding up and down the cylinder, a *reciprocating* motion.

There are still many problems, however. How can the up-and-down motion of the piston be converted into useful rotary motion? How can exhaust gases be removed? How can new fuel mixture be brought into the combustion chamber? Studying the engine's basic parts can help answer these questions.

BASIC ENGINE PARTS

The internal combustion engine parts discussed here are those of the four-stroke cycle gasoline engine, the most common type. The most essential of these parts are briefly listed.

- *Cylinder:* hollow; stationary; piston moves up and down within the cylinder.
- *Piston:* fits snugly into the cylinder, but still can be moved up and down.
- *Crankshaft:* converts reciprocating motion into more useful rotary motion.
- *Connecting rod:* connects the piston and the crankshaft.
- *Valves:* "doors" for admitting fuel mixture and releasing exhaust.
- *Crankcase:* the body of the engine; it contains most of the engine's moving parts.

See figure 3-1 for illustrations of the basic parts of two- and four-cycle engines. Figures 3-2 and 3-3, page 20, show cross sections of two small gas engines: outboard motor and lawn mower engine.

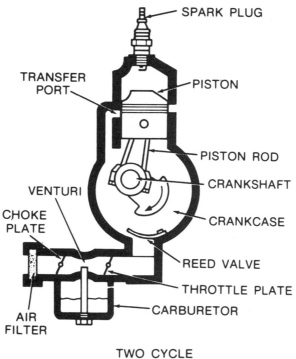

TWO CYCLE

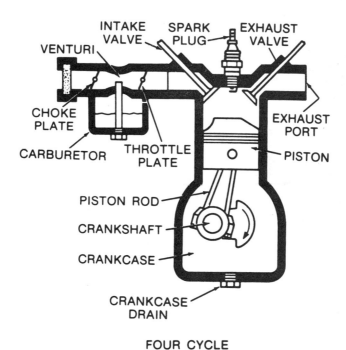

FOUR CYCLE

Fig. 3-1 Basic parts of two- and four-cycle engines

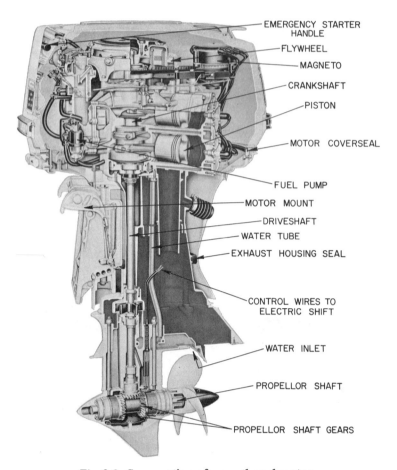

Fig. 3-2 Cross section of an outboard motor

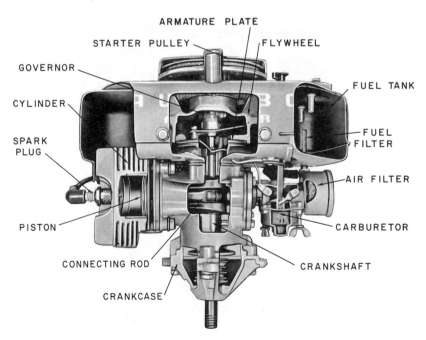

Fig. 3-3 Cross section of a lawn mower engine

CYLINDER

The cylinder is a finely machined part which is usually cast as a part of the crankcase. It can be thought of as the upper part of the crankcase. The piston slides up and down in the cylinder. The fit is close; the piston is only a few thousandths of an inch smaller than the cylinder. Although an engine may be basically made from aluminum, the cylinder itself is usually made of cast iron. Normally the aluminum body of the engine is cast around the cylinder and the two are inseparable. On some engines the cylinder section can be removed—a cylinder liner. This is particularly true of very large engines. Light-duty engines often have aluminum cylinders. In engine specifications the diameter of the cylinder is an important measurement. It is referred to as the *bore* of the engine. Figure 3-4, shows four different cylinder arrangements.

CYLINDER HEAD

The cylinder head forms the top of the combustion chamber and is exposed to great heat and pressure. This is where the action is—the few cubic inches made up by the top of the piston, the upper cylinder walls and the cylinder head. The cylinder head must be tightly bolted to the cylinder. It is important to tighten all cylinder head bolts with an even pressure and in their correct order (see figure 3-5, page 22), so that uneven stress is not set up in the cylinder walls. Such stress can distort or warp the cylinder walls.

A gasket between the two metal surfaces makes an airtight seal, figure 3-6, page 22. Gaskets are made of relatively soft material such as fiber, soft metal, cork, and rubber. Cylinder head gaskets must be able to hold the high pressure of combustion without blowing out and they must be heat resistant. The cylinder head also contains the threaded hole for the spark plug.

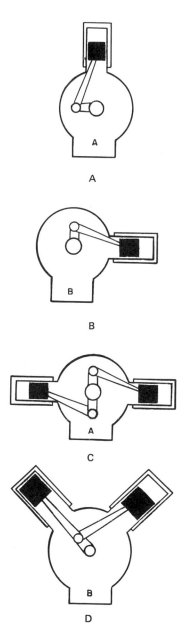

Fig. 3-4 Cylinder arrangements: A)vertical; B)horizontal; C)opposed; and D)V-type

There are many engines that have the cylinder and the cylinder head cast as one piece, therefore there are no gaskets or head bolts. A threaded spark plug hole is still located at the top of the cylinder head. Many small two-cycle engines use this design.

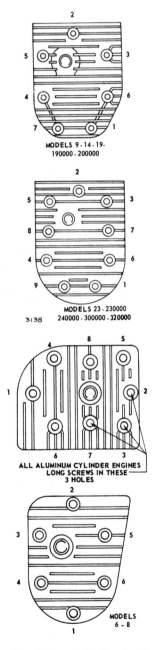

Fig. 3-5 Cylinder head. Note the sequence for tightening the head bolts.

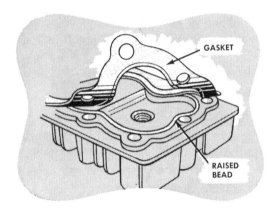

Fig. 3-6 The cylinder head gasket provides an airtight seal.

PISTON

The piston slides up and down in the cylinder. It is the only part of the combustion chamber that can move when the pressure of the rapidly burning fuel mixture is applied. The piston can be made of cast iron, steel, or aluminum; aluminum is commonly used for small engines because of its light weight and its ability to conduct heat away rapidly.

There must be clearance between the cylinder wall and the piston to prevent excessive wear. The piston is 0.003 to 0.004 of an inch smaller than the cylinder. To check this clearance, a special feeler gauge is necessary.

The piston top, or crown, may be flat, convex, concave, or many other shapes. Manufacturers select the shape that causes turbulence of the fuel mixture in the combustion chamber, and promotes smooth burning of the fuel mixture. The piston has machined grooves near the top to accommodate the piston rings (see figure 3-7, page 23). Some pistons are cam ground—that is, not perfectly round—to provide for expansion.

The distance the piston moves up and down is called the *stroke* of the engine. The uppermost point of the piston's travel is called its *top dead center* position (TDC). The lowest point of the piston's travel is called its *bottom dead center* (BDC). Therefore, stroke is the distance between TDC and BDC.

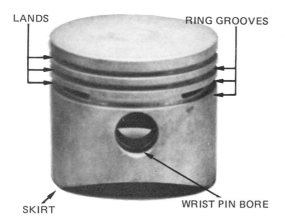

Fig. 3-7 The parts of a piston

PISTON RINGS

Piston rings provide a tight seal between the piston and the cylinder wall. Without the piston rings much of the force of combustion would escape between the piston and cylinder into the crankcase. The piston rings mainly act as a *power seal,* ensuring that a maximum amount of the combustion pressure is used in forcing the piston down the cylinder. Since the piston rings are between the piston and cylinder wall, there is a small area of metal sliding against the cylinder wall. Therefore, the piston rings reduce friction and the accompanying heat and wear that are caused by friction. Another function of the piston ring is to control the lubrication of the cylinder wall.

The job of the piston rings might appear simple but the piston rings may have to work under harsh conditions such as (1) distorted cylinder walls, due to improper tightening of head bolts, (2) cylinder out of round, (3) worn or scored cylinder walls, (4) worn piston, and (5) conditions of expansion due to intense heat. The important job of providing a power

seal can become difficult if the engine is worn or has been abused.

Piston rings are made of cast iron or steel (chrome is also used) and are finely machined. Sometimes their surfaces are plated with other metals to improve their action. There are many designs of piston rings and some rings, especially oil rings, may consist of more than one piece.

Two serious problems, blow-by and oil pumping, can be caused by bad piston rings, bad piston, warped cylinder, distortion, and/or scoring. In blow-by, the pressures of combustion are great enough to break the oil seal provided by the piston rings. When this happens, the gases of combustion force their way into the crankcase. The result is a loss in engine power and contamination of the crankcase.

Piston rings that are not fitted properly may start pumping oil. This pumping action leads to fouling in the combustion chamber, excessive oil consumption, and poor combustion characteristics. The rings have a side clearance which can cause a pumping action as the piston moves up and down.

A piston usually has three or four piston rings. Some have only one ring. The top piston ring is a compression ring, figure 3-8, which exerts a pressure of eight to twelve pounds on the cylinder wall. The ring has a clearance in its groove: it can move slightly up and down and expand slightly in and out. The second ring from the top is also a compression ring. These two rings form the power seal.

Fig. 3-8 Compression ring

Fig. 3-9 Oil control ring

The third ring down (and fourth if one is present), is an oil control ring, figure 3-9. The job of this piston ring is to control the lubrication of the cylinder wall. These rings spread the correct amount of oil on the cylinder wall, scraping the excess from the wall and returning it to the crankcase. The rings are slotted and grooves are cut into the piston behind the ring. This enables much of the excess oil to be scraped through the piston and it then drips back into the crankcase. It should be noted that compression rings have a secondary function of oil control.

Two clearances (or tolerances) that are important in piston rings are end gap and side clearance. The end gap is the space between the ends of the piston ring, measured when the ring is in the cylinder. The gap must be large enough to allow for expansion due to heat but not so large that power loss due to blow-by will result. Often 0.004 inch is allowed for each inch of piston diameter (top ring), and 0.003 inch is allowed for rings under the top ring. Since the second and third rings are exposed to less heat their allowance can be smaller.

Side clearance, figure 3-10, or ring groove clearance is also provided for heat expansion. Often 0.0025 inch is allowed for the top ring and 0.003 inch is allowed on the second and third rings.

PISTON PIN OR WRIST PIN

The piston pin, figure 3-11, is a precision ground steel pin that connects the piston and the connecting rod. This pin can be solid or hollow; the hollow piston pin has the advantage of being lighter in weight. Because piston pins are subjected to heavy shocks when combustion takes place, high-tensile strength steel is used. To keep noise at a minimum, the piston pin fits to a very close tolerance in the connecting rod. The piston pin does not rotate. It has a rocking motion similar to the action of a person's wrist when holding a bar tightly and swinging it back and forth, hence

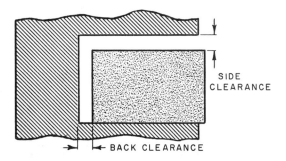

Fig. 3-10 Cross section showing a piston ring in its groove. Notice back clearance and side clearance.

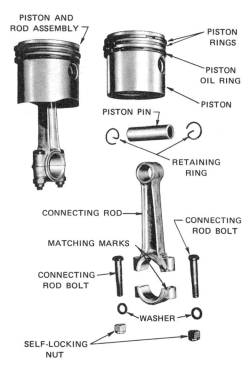

Fig. 3-11 Piston, piston pin, connecting rod, and associated parts

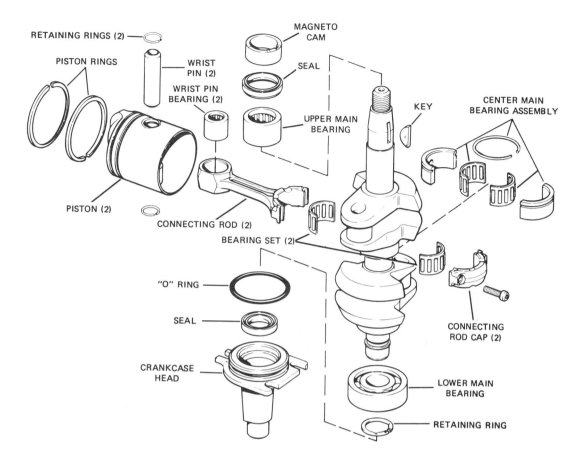

Fig. 3-12 Crankshaft, pistons, and connecting rod components—two-cycle, two-cylinder engine (note crankshaft)

the name *wrist pin.* Heavy shocks, close tolerances, and the rocking motion make the lubrication of this part difficult. In most engines, lubricating oil is squirted or splashed on it.

CONNECTING ROD

The connecting rod connects the piston (with the aid of the piston pin) and the crankshaft, figure 3-12. Many small engines have cast aluminum connecting rods; larger engines can have steel connecting rods. Generally the cross section of the connecting rod is an I beam shape. The lower end of the connecting rod is fitted with a cap. Both the lower end and the cap are accurately machined to form a perfect circle. This cap is bolted to the rod,

encircling the connecting rod bearing surface on the crankshaft. Many connecting rods are fitted with replaceable bearing surfaces called rod bearings or inserts, used especially on heavy-duty or more expensive engines.

CRANKSHAFT

The crankshaft is the vital part that converts the reciprocating motion of the piston into rotary motion. One end of the crankshaft has a provision for power takeoff (connecting accessory to use engine's power) and the other is machined to accept the flywheel.

Crankshafts are carefully forged and machined steel. Their bearing surfaces are large because they must accept a great amount

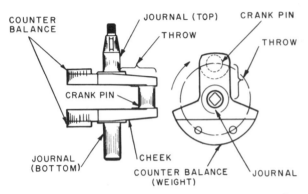

Fig. 3-13 Single-throw crankshaft: one crank pin for one cylinder

of force. Since they revolve at high speeds, they must be well balanced. Figure 3-13 is an example of a single-throw crankshaft.

Figure 3-14 shows two types of small engine designs. Engines designed with *vertical* crankshafts, figure 3-15, page 27, are excellent for applications such as rotary power lawn mowers where the blade is bolted directly to the end of the crankshaft. Most wheeled applications such as motorbikes and go-carts use *horizontal* crankshaft engines which enable the power to be delivered to an axle through belts or chains. Multiposition engines such as chain saws are another variation. The crankshafts themselves are much the same but other changes in engine design are necessary to meet the requirements of crankshaft position. An engine can have clockwise or counterclockwise crankshaft rotation, depending upon its application and design.

CRANKCASE

The crankcase is the body of the engine. It houses the crankshaft and has bearing surfaces on which the crankshaft revolves. Many other parts are located within the crankcase; connecting rods, cam gear, camshaft, and lubrication mechanism. The cylinder and bottom of the piston are exposed to the crankcase. On four-cycle engines a reservoir of oil is found within the crankcase. This lubricating

Fig. 3-14 Two types of small engine designs

oil is splashed, pumped, or squirted onto all the moving parts located in the crankcase. Figure 3-16, page 27, shows the components of a short block assembly, and

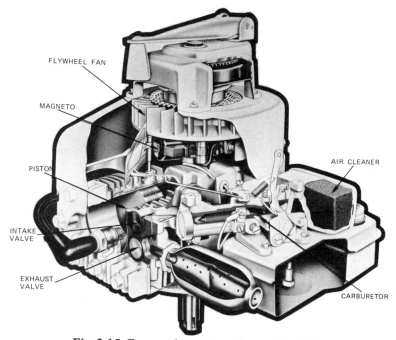

FLYWHEEL FAN

MAGNETO

PISTON

AIR CLEANER

INTAKE
VALVE

EXHAUST
VALVE

CARBURETOR

Fig. 3-15 Four-cycle engine with vertical shaft

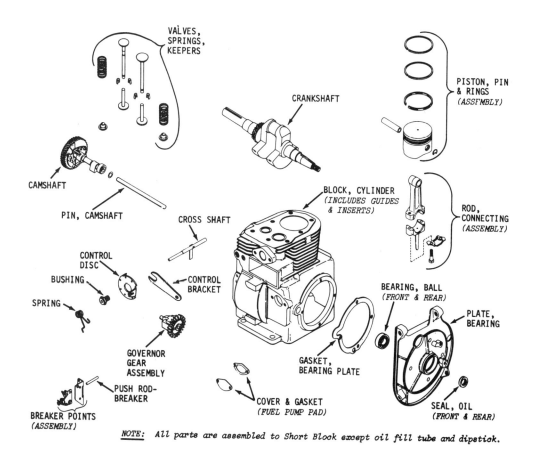

VALVES,
SPRINGS,
KEEPERS

PISTON, PIN
& RINGS
(ASSEMBLY)

CRANKSHAFT

CAMSHAFT

BLOCK, CYLINDER
(INCLUDES GUIDES
& INSERTS)

ROD,
CONNECTING
(ASSEMBLY)

PIN, CAMSHAFT

CROSS SHAFT

CONTROL
DISC

BUSHING

CONTROL
BRACKET

SPRING

BEARING, BALL
(FRONT & REAR)

PLATE,
BEARING

GOVERNOR
GEAR
ASSEMBLY

GASKET,
BEARING PLATE

PUSH ROD-
BREAKER

BREAKER POINTS
(ASSEMBLY)

COVER & GASKET
(FUEL PUMP PAD)

SEAL, OIL
(FRONT & REAR)

<u>NOTE:</u> All parts are assembled to Short Block except oil fill tube and dipstick.

Fig. 3-16 Components of a short block assembly

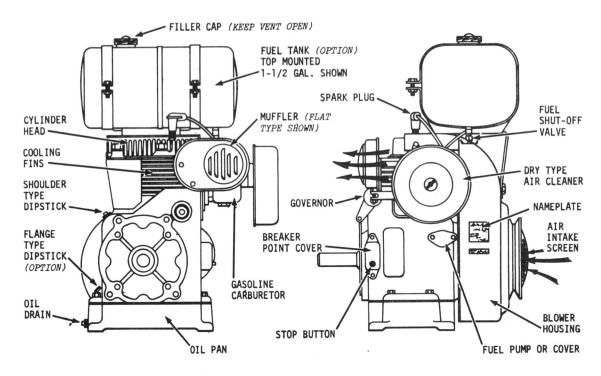

Fig. 3-17 Single-cylinder engine

a single-cylinder engine is illustrated in figure 3-17.

FLYWHEEL

The flywheel is mounted on the tapered end of the crankshaft. A key keeps the two parts solidly together, figure 3-18, and they revolve as one part. The flywheel is found on all small engines. It is relatively heavy and helps to smooth out the operation of the engine. That is, after the force of combustion has pushed the piston down, the momentum of the flywheel helps to move the piston back up the cylinder. This tends to minimize or eliminate sudden jolts of power during combustion. The more cylinders the engine has, the less important this smoothing out action becomes.

The flywheel can also be a part of the engine's cooling system by having air vanes to scoop air as the flywheel revolves. This air

is channeled across the hot engine by the cover or shroud, to carry away heat.

Most flywheels on small gasoline engines are part of the ignition systems; they have permanent magnets mounted in them. These permanent magnets are an essential part of the magneto ignition system.

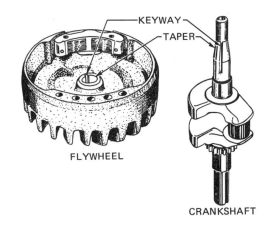

Fig. 3-18 A key locks the flywheel and crankshaft together.

VALVES

Figure 3-19 shows the parts of a valve. The most common valves are called *poppet valves.* With these valves, exhaust gases can be removed from the combustion chamber and new fuel mixture can be brought in. To accomplish this, the valve actually pops open and snaps closed. In the normal four-cycle combustion chamber there are two valves, one to let the exhaust out (exhaust valve) and one to allow the fuel mixture to come in (intake valve).

The two valves may look almost identical, but they are different. The exhaust valve must be designed to take an extreme amount of punishment and still function perfectly. Not only is it subjected to the normal heat of combustion (about 4500° F), but when it opens, the hot exhaust gases rush by it. When the valve returns to its seat, a small area touches, making it difficult for the valve's heat to be conducted away. This rugged valve opens and closes 1800 times a minute if the engine is operating at 3600 RPM. Exhaust valves are made from special heat-resisting alloys.

Valves can be made to rotate a bit each time they open and close. This action tends to prevent deposits from building up at the valve seat and margin, and in addition provides for even wear and exposure to temperatures. This provision greatly increases the life of the valves, figure 3-20.

Intake valves operate in the same manner as exhaust valves except they are not subjected to the extreme heat conditions. Each time they open new fuel mixture rushes by, helping to cool the intake valve.

Operation of the valve requires the help of associated parts—the valve train and camshaft, figure 3-21. The valve train consists of the valve (intake or exhaust), valve spring, tappet or valve lifter, and cam. The valve spring closes the valve, holding it tightly

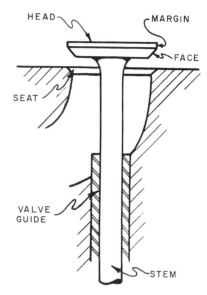

Fig. 3-19 A valve

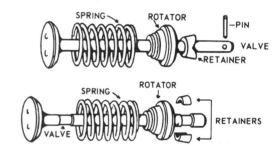

Fig. 3-20 "Rotocap" provides for even wear and long valve life

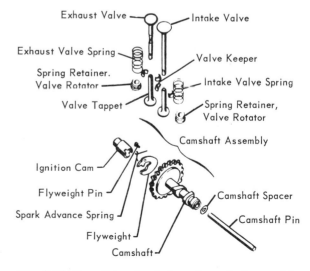

Fig. 3-21 Complete valve train and camshaft assembly

against its seat. The tappet rides on the cam and there is a small clearance between the tappet and the valve stem. Some tappet clearances are adjustable especially on larger and more expensive engines. The cam or lobe pushes up on the tappet; the tappet pushes up on the valve stem; and the valve opens when the high spot on the cam is reached. The opening and closing of the valves must be carefully timed in order to get the exhaust out at the correct time and the new fuel mixture in at the correct time.

VALVE ARRANGEMENT

Engines may be designed and built with one of several valve arrangements.

- *The L-head:* both valves open up on one side of the cylinder. This is the most common arrangement in small gasoline engines. See figures 3-22 and 3-23.

- *The I-head:* both valves open down over the cylinder. See figure 3-24.

- *The F-head:* one valve opens up and one valve opens down; both are on the same side of the cylinder. See figure 3-25.

- *The T-head:* one valve is on one side of the cylinder and the other valve is on the opposite side; both valves open up. See figure 3-26.

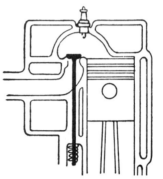 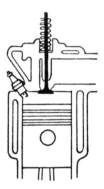

Fig. 3-23 L-head valve arrangement

Fig. 3-24 I-head valve arrangement

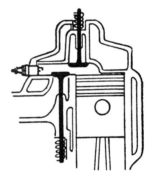

Fig. 3-25 F-head valve arrangement

Fig. 3-22 L-head design used for most four-stroke cycle small engines

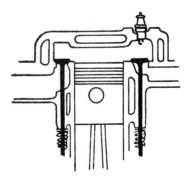

Fig. 3-26 T-head valve arrangement

CAMSHAFT

The camshaft and its associated parts control the opening and closing of the valves. The cams on this shaft are egg-shaped. As they revolve, their noses cause the valves to open and close. A tappet rides on each cam. When the nose of the cam comes under the tappet the valve is pushed open, admitting new fuel mixture or allowing exhaust to escape, depending on which valve is opened.

The camshaft is driven by the crankshaft. On four-cycle engines the camshaft operates at one-half the crankshaft speed.

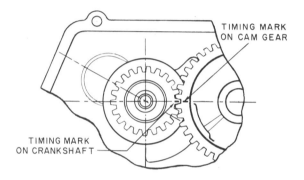

Fig. 3-27 Aligning timing marks on the crankshaft gear and camshaft gear. Notice the larger camshaft gear revolves at one-half crankshaft speed.

(If the engine is operating at 3400 RPM, the camshaft revolves at 1700 RPM.) The crankshaft and camshaft must be in perfect synchronization, so most engines have timing marks on the gears, figure 3-27, to ensure the correct reassembly of the gears.

FOUR-STROKE CYCLE OPERATION

All of these parts and many others must be assembled to operate the gasoline engine. The most common method of engine operation is the four-stroke cycle. Most small gasoline engines operate on this principle, as do automobile engines. The four-stroke cycle engine is an entirely successful producer of power.

Four-stroke cycle, figure 3-28, means that it takes four strokes of the piston to complete the operating cycle of the engine. The piston goes down the cylinder, up, down, and back up again to complete the cycle. This takes two revolutions of the crankshaft; each stroke is one-half revolution.

When the piston travels down the cylinder, this is intake; fuel mixture enters the combustion chamber. Reaching the bottom of its stroke, the piston comes up on compression and the fuel mixture is pressed into a

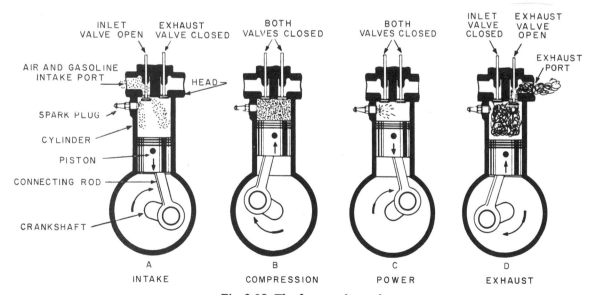

Fig. 3-28 The four-stroke cycle

small space at the top of the combustion chamber. Power comes when the fuel mixture is ignited, pushing the piston back down. The piston moves up once more pushing the exhaust gases out of the combustion chamber. This is exhaust. These four strokes complete the engine's cycle and require two revolutions of the crankshaft. As soon as one cycle is complete, another begins. If an engine is operating at a speed of 3600 RPM there are 1800 cycles each minute.

Intake Stroke

If the fuel mixture is to enter the combustion chamber, the intake valve must be open and the exhaust valve must be closed. With the intake valve open and the piston traveling down the cylinder, the fuel mixture rushes in easily. This is the *intake stroke.*

Pushing down on everything is an atmospheric pressure of 14.7 pounds per square inch (psi) at sea level. When the piston is at the top of its stroke there is normal atmospheric pressure in the combustion chamber. When the piston travels down the cylinder there is more space for the same amount of air and the air pressure is reduced; a partial vacuum is created. The intake valve opens, and the normal atmospheric pressure rushes in to equalize this lower air pressure in the combustion chamber. The fuel mixture is actually pushed into the combustion chamber, even though it seems to be sucked in by the partial vacuum.

Compression Stroke

When the piston reaches the bottom of its stroke and the cylinder is filled with fuel mixture, the intake valve closes and the exhaust valve remains closed. The piston now travels up the cylinder compressing the fuel mixture into a smaller and smaller space. When the fuel mixture is compressed into this small space it can be ignited more easily and it expands very rapidly.

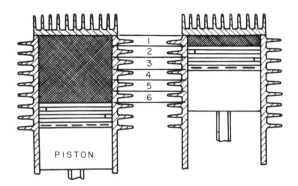

Fig. 3-29 An engine with a six-to-one compression ratio

The term *compression ratio* applies to this stroke. If the piston compresses the fuel mixture into one-sixth of the original space the compression ratio is 6:1, figure 3-29. An engine whose piston compresses the fuel mixture into one-eighth the original space has a compresssion ratio of 8:1. The higher the compression, the more powerful the force of combustion is.

Power Stroke

When the piston reaches the top of its stroke, and the fuel mixture is compressed, a spark jumps across the spark plug igniting the fuel mixture. The fuel mixture burns very rapidly, almost exploding, and the burning, expanding gases exert a great pressure in the combustion chamber. The combustion pressure is felt in all directions, but only the piston is free to react. It reacts by being pushed rapidly down the cylinder. This is the *power stroke:* rapidly burning fuel creates a pressure that pushes the piston down, turning the crankshaft and producing usable rotary motion.

Both intake and exhaust valves must be closed for this stroke. The force of combustion must not leak out the valves. Also, this force should not blow-by the piston rings. Correctly fitting piston rings prevent this. All gaskets, namely the cylinder head gasket and spark plug gasket, should be airtight.

The power must be transmitted to the top of the piston and not lost elsewhere.

Exhaust Stroke

When the piston reaches the bottom of its power stroke the momentum of the flywheel and crankshaft bring the piston back up the cylinder. This is the *exhaust stroke.* Now the exhaust valve opens, the intake valve remains closed, and the piston pushes the burned exhaust gases out of the cylinder and combustion chamber.

TWO-STROKE CYCLE OPERATION

Another common operating principle for gasoline engines is the two-stroke cycle. This engine, illustrated in figure 3-30, is also entirely successful and in common use today, especially for outboard motors, chain saws, and many lawn mowers. The two-stroke cycle engine is constructed somewhat differently from the four-stroke cycle engine. See figure 3-31 for a comparison of moving parts.

Two-stroke cycle means that it takes two strokes of the piston to complete the operating

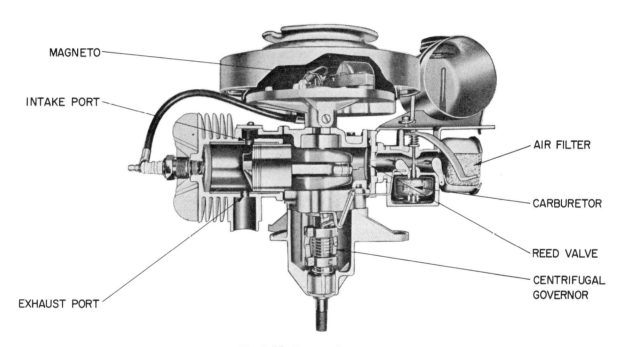

Fig. 3-30 Two-cycle engine

MOVING PARTS—4-CYCLE ENGINE

MOVING PARTS—2-CYCLE ENGINE

Fig. 3-31 Comparison of four-cycle and two-cycle engines

cycle of the engine. See figure 3-32. The piston goes down the cylinder and back up again and the cycle is completed. This takes only one revolution of the crankshaft.

With the piston at the top of its stroke, and the fuel mixture tightly compressed in the combustion chamber, the engine is ready for its first stroke, the *power stroke.* A spark ignites the fuel mixture and the great pressure pushes the piston rapidly down the cylinder. As the piston nears the bottom of its stroke, the exhaust ports begin to uncover and the exhaust gases, which are still hot and under pressure, start to rush through the ports and out of the engine. Just after the exhaust ports begin to uncover, the piston uncovers the intake ports. Fuel mixture rushes into the cylinder, being deflected to the top of the cylinder first and finally scavenging out the last bit of exhaust. By this time the piston has reached the bottom of its stroke and is on its way back up the cylinder; the ports are sealed off, trapping the fuel mixture. This stroke is *compression,* the piston pushing the fuel mixture into a smaller and smaller space. The two strokes, then, are power and compression, with intake and exhaust taking place between the two.

One question remains to be answered: why does the fuel mixture rush through the intake ports into the combustion chamber? It is not because there is a partial vacuum in the combustion chamber as in the four-stroke cycle engine. With the exhaust ports open, the combustion chamber soon would be at atmospheric pressure. The answer lies in the crankcase and reed valves. As the piston moves up the cylinder, a low pressure, or partial vacuum, is created in the crankcase; there is a larger space for the same amount of air. The greater atmospheric pressure outside rushes through the carburetor, pushing open the springy reed valves, filling the crankcase with fuel mixture. When the pressure in the crankcase and the atmospheric pressure are just about equal, the leaf valves spring shut.

As the piston moves back down the cylinder on its power stroke, the fuel mixture, trapped in the crankcase, is put under a slight pressure. The leaf valve can open in only the opposite direction. When the piston nears the bottom of its stroke, the exhaust ports uncover allowing the exhaust to escape. The intake ports open slightly after the exhaust ports and the pressure in the crankcase pushes through the intake ports and into the cylinder, filling it with new fuel mixture.

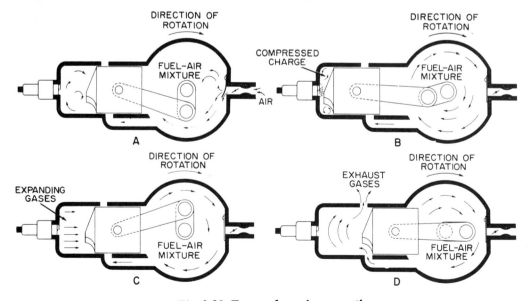

Fig. 3-32 Two-cycle engine operation

The cycle is then repeated. If the engine operates at 4000 RPM the cycle is completed 4000 times each minute. There is a power stroke for each revolution of the crankshaft.

CRANKCASE

The crankcase of a two-stroke cycle engine is designed to be as small in volume as possible. The smaller the volume, the greater the presure created as the piston comes back down the cylinder. The greater the crankcase pressure, the more efficient the transfer of fuel from the crankcase through the transfer ports into the cylinder.

INTAKE AND EXHAUST PORTS

Intake and exhaust ports are holes drilled in the cylinder wall which allow exhaust gases to escape and a new fuel mixture to enter. They are located near the bottom of the cylinder and are covered and uncovered by the piston. By using ports, many parts seen in the four-cycle engine are eliminated: valves, tappets, valve springs, cam gear, and camshaft. The remaining major moving parts are the piston, connecting rod, and crankshaft.

PISTON (CROSS SCAVENGED)

The piston still serves the same function in the cylinder, but for two-cycle engines the top is usually designed differently, figure 3-33.

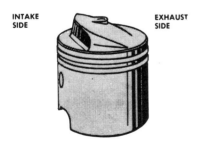

INTAKE SIDE   EXHAUST SIDE

Fig. 3-33 A piston for a two-cycle engine showing the exhaust side and the intake side

On the intake side there is a sharp deflection that sends the incoming fuel mixture to the top of the cylinder. On the exhaust side there is a gentle slope so that exhaust gases have a clear path to escape. The loop scavenged two-cycle engine does, however, have a flat top piston; it is discussed in a later section.

REED OR LEAF VALVES

These valves are located between the carburetor and crankcase. The valve itself is a thin sheet of springy alloy steel. It springs open to allow fuel mixture to enter the crankcase, and springs closed to seal the crankcase. There may be only one reed valve or there may be several reed valves working together. The majority of small gasoline engines operating on the two-cycle principle use reed valves even though it is possible to use a poppet valve in the crankcase. Several kinds of reed plates are shown in figure 3-34.

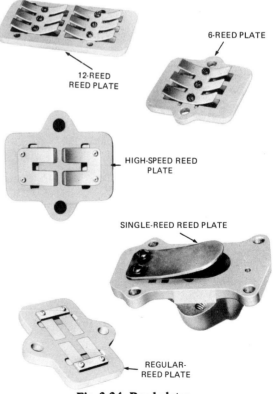

6-REED PLATE

12-REED REED PLATE

HIGH-SPEED REED PLATE

SINGLE-REED REED PLATE

REGULAR-REED PLATE

Fig. 3-34 Reed plates

OTHER TWO-CYCLE OPERATION

There are other methods of two-cycle operation; not all two-cycle engines use reed or leaf valves. The other engines also have the two strokes, compression and power, with intake and exhaust taking place between the two, but different valves are used to achieve the result.

ROTARY VALVE

Some two-cycle engines use a rotary valve, figure 3-35, for admitting fuel mixture. The rotary valve is a flat disc, with a section removed, that is fastened to the crankshaft. It normally seals the crankcase, but as the piston nears the top of its stroke and there is a slight vacuum in the crankcase, the valve is rotated to its open position. This allows fuel mixture to travel from the carburetor through the open valve and into the crankcase.

THIRD PORT DESIGN

This design has the regular exhaust ports and intake ports on the cylinder wall but in ad-

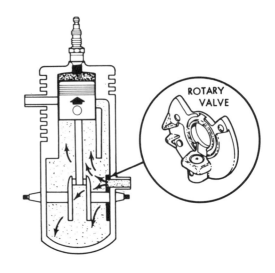

Fig. 3-35 Rotary valve

dition it has a third port in the cylinder wall (see figure 3-36). This third port is for admitting fuel mixture to the crankcase. The bottom of the piston skirt uncovers this port as the piston nears the top of its stroke. Naturally, the piston's upward travel creates a low pressure in the crankcase. When the port is uncovered, fuel mixture rushes into the crankcase. As the piston comes down on the power stroke

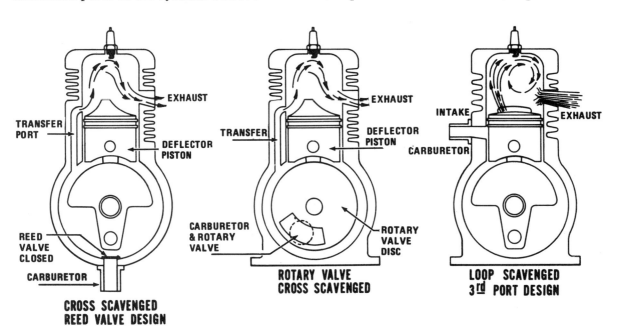

Fig. 3-36 Basic designs of two-cycle spark ignition engines

the port is sealed off and the trapped mixture is pressurized—ready for transfer through the intake ports into the cylinder.

POPPET VALVES

There are some two-cycle engines that use poppet valves to admit new fuel mixture into the crankcase. These poppet valves may be spring loaded and operated by differences in crankcase pressure. They open when the partial vacuum in the crankcase overcomes a slight spring tension. This happens when the piston is on its upward stroke. With the poppet valve open, fuel mixture rushes into the crankcase. The poppet valve may also be operated by cam action. In this case a crankshaft cam opens the valve for the piston's upward stroke.

LOOP SCAVENGING

The loop scavenged two-cycle engine is basically the same as any other two-cycle engine. However, it does not need the contoured piston of the cross scavenged design. Any type of valve arrangement can be used for admitting the fuel into the crankcase. The difference is that the intake ports are drilled into the cylinder walls at an angle, aiming the incoming fuel mixture toward the top of the cylinder. Usually there are two pairs of intake ports angling in from the sides, directing the fuel mixture to the top of the cylinder, around and back through the center of the cylinder to scavenge out the last of the exhaust through the exhaust ports. Loop scavenging provides a more complete removal of exhaust gases and produces somewhat more horsepower per unit weight.

GENERAL REVIEW QUESTIONS

1. Explain the term internal combustion.

2. List the most essential parts of the engine.

3. Of what metal is the cylinder made?

4. What purpose does a gasket serve?

5. What is a combustion chamber?

6. What purpose do piston rings serve?

7. What is the function of the crankshaft?

8. 'What does the flywheel accomplish?

9. Explain how the valves are operated.

10. At what speed does the camshaft revolve relative to the crankshaft?

FOUR-STROKE CYCLE REVIEW QUESTIONS

1. Explain what causes the fuel mixture to rush into the cylinder during the intake stroke.

2. What is accomplished on the compression stroke?

3. Explain the power stroke.

4. How many revolutions are required for the complete cycle?

5. Explain compression ratio.

TWO-STROKE CYCLE REVIEW QUESTIONS

1. What parts are replaced by the intake and exhaust ports?

2. How are most two-stroke cycle pistons different from four-stroke cycle pistons?

3. Explain the action of the reed or leaf valves.

4. Explain how intake and exhaust take place on a two-cycle engine.

5. What part does the crankcase play in regard to the fuel mixture?

6. How many revolutions are required for the complete cycle?

CLASS DISCUSSION TOPICS

- What other systems does an engine need in addition to the basic moving parts?
- Discuss the basic differences in two- and four-cycle engines.
- Discuss the advantages and disadvantages of two- and four-cycle engines.
- Discuss how parts can be damaged by careless handling.

CLASS DEMONSTRATION TOPICS

- Inspect the basic parts of the two-cycle and four-cycle engine.
- Remove the cylinder head of a four-cycle engine, rotate the crankshaft, and observe the action of the valves.
- Remove the valve spring cover of a four-cycle engine, rotate the crankshaft, and observe the action of the valve lifters and valve springs.
- Remove the carburetor of a two-cycle engine, turn the engine over rapidly, and observe the action of the leaf valves. Note: This action is not pronounced and must be carefully observed.
- Using a disassembled engine, connect the piston, connecting rod, and crankshaft to illustrate how reciprocating motion is converted to rotary motion.
- Using a camshaft and crankshaft, point out timing marks.
- Using a camshaft and crankshaft, count the gear teeth to illustrate the speed relationship of the two gears.

LABORATORY EXPERIENCE 3-1
ENGINE DISASSEMBLY (FOUR-STROKE CYCLE)

Note: Laboratory Experiences 3-1 and 3-2 are well suited for a class demonstration, especially if the students are beginners. If, however, students have a background in engine disassembly, it can be accomplished at the end of unit 3 in the text. Another alternative is for the instructor to demonstrate the work with unit 3 and then have the students do Laboratory Experiences 3-1 and 3-2 while they study unit 9 on simple repairs.

In this exercise, a four-stroke cycle engine is disassembled, the basic parts are studied, and then the engine is reassembled. It is important to be able to identify these parts and know how they fit and work together. Speed in engine disassembly is not the important thing. What is important is a careful, serious approach to the job. Students are to record their work after each step of the disassembly procedure is completed.

Disassembly Procedure

Instructors may want to supplement or revise specific steps of this procedure since there are many makes of engines. The following disassembly procedure is a general guide.

1. Drain the oil from the crankcase.

2. Disconnect the spark plug cable and remove the spark plug.

3. Drain the gas tank, disconnect the fuel line, and remove the gas tank.

4. Remove the air cleaner. (Use care if it contains oil.)

5. Drain the carburetor, remove throttle and governor connections, and remove the carburetor from the engine.

6. Remove metal air shrouding.

7. Remove the crankshaft nut; also remove grass screens, starter pulleys, etc.

8. Remove the flywheel.

9. Remove the magneto plate assembly. (The main bearing is on the flywheel side of this plate.)

10. Remove the cylinder head.

11. Remove the crankcase from the engine base. (This may not be necessary if the engine has an inspection plate.)

12. Unbolt the connecting rod cap and push the piston and rod up and out of the cylinder.

13. Remove the crankshaft by pulling it out. (The main bearing plate may have to be loosened to do this.)

Part	Disassembly (nuts, bolts, etc.)	Operation performed	Tool used

Reassembly Procedure

Reverse the disassembly procedure. Listed are some points to remember.

1. Tighten all machine screws and bolts securely.
2. Be certain the piston is reinstalled exactly as it came out.
3. Line up the timing marks on the crankshaft with the timing mark on the camshaft.
4. Line up the match mark on the connecting rod cap with the match mark on the connecting rod.
5. Be certain the flywheel keyway slips into the key on the crankshaft.
6. Check all gaskets; replacement may be necessary, especially if the engine is to be operated.
7. Refill the crankcase with oil if the engine is to be operated.
8. Refill the gas tank if the engine is to be operated.

REVIEW QUESTIONS

Study the basic parts of the engine and answer the following questions.

1. Why should most gaskets be replaced when the engine is reassembled?

2. Explain the value of match marks on the connecting rod and cap.

3. Why is uniform tightening of cylinder head bolts important?

4. Explain why timing marks are used on the crankshaft and camshaft.

5. How many piston rings are on the piston? What type are they?

LABORATORY EXPERIENCE 3-2
VALVE TRAIN AND CAMSHAFT (FOUR-STROKE CYCLE)

Note: This laboratory experience may be done at the completion of disassembly on Laboratory Experience 3-1 or it may be done starting with a fully assembled engine. If students start with a fully assembled engine they should follow the disassembly procedure for Laboratory Experience 3-1, then continue on with the disassembly of the valve train and camshaft.

The valves, valve springs, and valve spring retaining parts are removed. Then the camshaft and valve lifters are removed. These parts are studied and then reassembled. Students are to record their work after each step of the disassembly procedure is completed.

Disassembly Procedure

Instructors may want to supplement or revise specific steps of this procedure since there are many makes of engines. The following disassembly procedure is a general guide.

1. Remove the valve plate exposing the valve springs, valve lifters, and valve stems.

2. Compress the valve spring with a valve spring compressor.

3. Remove the valve spring retainers, figure 3-37, by slipping or flipping them out.

4. Pull the valve out of the engine.

5. Remove the valve spring, still compressed.

6. Remove the camshaft. If the camshaft is held in the crankcase with a camshaft support pin it must be driven out with a blunt punch. In most cases, drive the punch from the takeoff side toward the flywheel side.

   NOTE: Other engines may be constructed with a one-piece camshaft that comes out when the crankshaft is removed.

7. Pull out the camshaft.

8. Remove the valve lifters; they will probably fall out when the camshaft is removed.

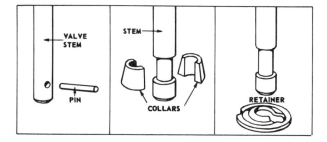

Fig. 3-37 Valve spring retainers

Part	Disassembly (nuts, bolts, etc.)	Operation performed	Tool used

Reassembly Procedure

Reverse the disassembly procedure. Listed are some points to remember.

1. Reinstall the intake and exhaust valves in their correct places.

2. On some engines a special tool is needed to aid in the installation of valve spring retainers.

3. Reinstall the camshaft support pin correctly: from the flywheel side to the power takeoff side, in most cases.

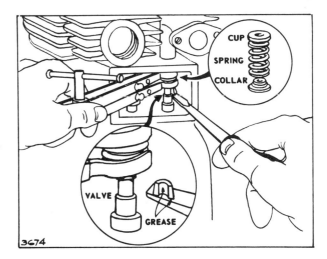

Fig. 3-38 One technique for reinstalling valve spring retainers

REVIEW QUESTIONS

Study the valves and associated parts and answer the following questions.

1. Why are exhaust valves and intake valves made differently?

2. Examine the two valves. How can they be told apart?

3. What is the function of the valve spring?

4. Why does the camshaft rotate at one-half the crankshaft speed?

5. How many degrees apart are the intake and exhaust cams? Why?

LABORATORY EXPERIENCE 3-3
EXHAUST AND INTAKE PORTS (TWO-STROKE CYCLE)

The exhaust baffle plate, muffler, or exhaust manifold (different terms, all relating to the same area) are removed. The ports are examined, then the engine is turned over and the ports are observed as they are covered and uncovered by the piston. Students are to record their work after each step of the disassembly procedure is completed.

Disassembly Procedure

Instructors may want to supplement or revise specific steps of this procedure since there are many makes of engines. The following disassembly procedure is a general guide.

1. Drain the gas tank.
2. Drain the carburetor.
3. Remove the spark plug cable from the spark plug.
4. Remove the spark plug from the cylinder.
5. Remove any metal shrouding that is around the exhaust area.
6. Remove the exhaust baffle plate, muffler, or exhaust manifold, thus exposing the exhaust ports.

Part	Disassembly (nuts, bolts, etc.)	Operation performed	Tool used

Reassembly Procedure

Reverse the disassembly procedure.

REVIEW QUESTIONS

Study the exhaust ports and their operation and answer the following questions.

1. If the engine has been in operation, were any carbon deposits observed around the exhaust ports? Describe these deposits.

2. Explain why the engine turned over easily when the spark plug was removed.

3. How many ports make up the exhaust port team?

4. Is the exhaust side of the piston a gentle slope or a sharp deflection? Why?

5. Is loop scavenging used on the engine?

LABORATORY EXPERIENCE 3-4
REED VALVES (TWO-STROKE CYCLE)

The reed valve plate is removed and the reed valves are studied. The parts removed are then reassembled. Students are to record their work after each step of the disassembly procedure is completed.

Observation of the opening and closing of the reed valves can usually be done by removing the carburetor, leaving the reed valve plate installed on the crankcase. Then, turn the engine over rapidly. The action of the reed valves is not too obvious; therefore, they must be watched very closely. If the spark plug is removed from the cylinder the engine can be turned over more easily.

Disassembly Procedure

Instructors may want to supplement or revise specific steps of this procedure since there are many makes of engines. The following disassembly procedure is a general guide.

1. Drain the gas tank.

2. Remove any metal shrouding from the carburetor area.

Part	Disassembly (nuts, bolts, etc.)	Operation performed	Tool used

3. Drain the carburetor.

4. Remove throttle and governor connections.

5. Remove the carburetor.

6. Remove the spark plug from the cylinder.

7. Rotate the flywheel rapidly and observe the reed valve action.

8. Remove the reed valve plate from the crankcase.

9. Remove individual reed valves for inspection only if the engine is not to be operated again. The reed valves can be damaged through unnecessary handling.

Reassembly Procedure

Reverse the disassembly procedure.

REVIEW QUESTIONS

Study the reed valves and answer the following questions.

1. The reed valves are between what two main parts?

2. What condition in the crankcase causes the reed valves to be pulled open?

3. What condition in the crankcase causes the reed valves to spring closed?

4. How many individual reed valves work together as a team on the engine?

5. Trace the path of the fuel mixture in a two-cycle engine, beginning with the gasoline tank and continuing to the combustion chamber.

UNIT 4 FUEL SYSTEMS, CARBURETION, AND GOVERNORS

OBJECTIVES

After completing this unit the student will be able to:

- identify the basic parts of the carburetor and explain the function of each part.

- adjust the carburetor for maximum power and efficiency.

- explain the basic construction of the diaphragm fuel pump and discuss fuel pump operation.

- identify the basic parts of an air vane governor and explain how they work together.

- adjust the governed speed and discuss the action of the governor on the throttle.

The fuel system must maintain a constant supply of gasoline for the engine. The carburetor must correctly mix the gasoline and air together to form a combustible mixture which burns rapidly when ignited in the combustion chamber.

A typical fuel system contains a *gasoline tank* (reservoir for gasoline); a *carburetor* (mixing device for gasoline); the *fuel line* (tubes made of rubber or copper through which the gasoline passes from the gas tank to the carburetor); and an *air cleaner*, figure 4-1, (device for filtering air brought into the carburetor). In addition, the system may have a *shutoff valve*, figure 4-2, (valve at the gas tank that can cut off the gasoline supply when the engine is not in use); a *fuel pump* (pump that supplies the carburetor with a constant supply of gasoline); a *sediment bowl*, figure 4-3, page 52, (small glass bowl attached to the fuel line where dirt and other foreign matter can settle out); and a *strainer* (fine screen in the gas tank to prevent leaves and dirt from entering the fuel line).

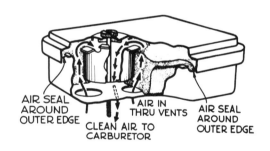

AIR SEAL AROUND OUTER EDGE   AIR IN THRU VENTS   AIR SEAL AROUND OUTER EDGE   CLEAN AIR TO CARBURETOR

Fig. 4-1 Air cleaner

Fig. 4-2 Fuel shutoff valve

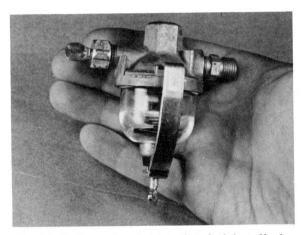

Fig. 4-3 Sediment bowl with built-in fuel shutoff valve

It should be noted that not all fuel systems contain all eight of the basic parts; shutoff valves, sediment bowls, fuel pumps, and air cleaners are not common to all engines.

A constant supply of gasoline must be available at the carburetor. To provide this supply, four methods are in common use today: suction, gravity, fuel pump, and pressurized tank.

SUCTION SYSTEM

The suction system, figure 4-4, is probably the simplest. With this method the gas

tank is located below the carburetor and the gasoline is sucked up into the carburetor. However, the gas tank cannot be very far away from the carburetor or the carburetor action will not be strong enough to pull the gasoline from the tank.

GRAVITY SYSTEM

With the gravity system, figure 4-5, the gas tank is located above the carburetor and the gasoline runs downhill to the carburetor. To prevent gasoline from continuously pouring through it, the carburetor has incorporated a *float* and *float chamber*. This float chamber provides a constant level of gasoline without flooding the carburetor. When gasoline is used, the float goes down, opening a valve to admit more gasoline to the float chamber. The float rises and shuts off the gasoline when it reaches its correct level.

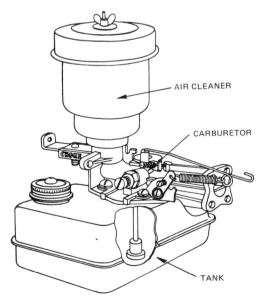

Fig. 4-4 Suction feed fuel system

Fig. 4-5 Gravity feed fuel system

In actual practice the float and float valve do not rapidly open and close, but assume a position allowing the correct amount of fuel to constantly enter the float chamber (see figure 4-6). If the engine is speeded up, a new position is assumed by the float and float valve, supplying an increased flow of gasoline.

FUEL PUMP

On many engines it is necessary to place the gas tank some distance from the carburetor. Therefore, the fuel must be brought to the carburetor by some means other than gravity or suction. One method is by using a fuel pump. The automobile engine uses a fuel pump. Among smaller engines, the outboard motor with a remote fuel tank is a good example of an engine that commonly uses a fuel pump.

The fuel pump used on many two-stroke cycle, outboard motors is quite simple, consisting of a chamber, inlet and discharge valve, a rubber diaphragm, and a spring. This fuel pump is operated by the crankcase pressure. As the piston goes up, low pressure in the crankcase pulls the diaphragm toward the crankcase, sucking gasoline through the inlet valve into the fuel chamber, figure, 4-7. As the piston comes down, pressure in the crank-

case pushes the diaphragm away from the crankcase. When this happens, the intake valve closes and the discharge valve opens, allowing the trapped fuel to be forced to the carburetor, figure 4-8.

PRESSURIZED FUEL SYSTEM

Another method for forcing fuel to travel long distances to the carburetor is the pressurized fuel system, figure 4-9, page 54. This method is sometimes used with outboard motors. Engines using this system have two hoses or lines between the gas tank and the engine: one brings gasoline to the carburetor; the other brings air, under pressure, from the

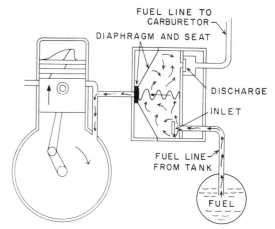

Fig. 4-7 Low pressure in the crankcase allows the fuel chamber to be filled.

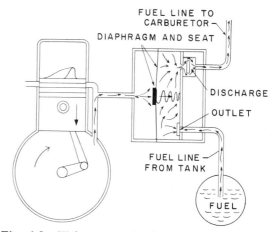

Fig. 4-8 High pressure in the crankcase forces the trapped fuel on to the carburetor.

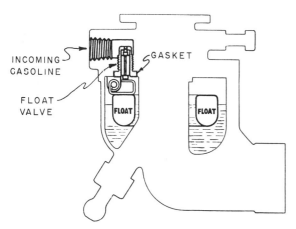

Fig. 4-6 Cutaway of a carburetor, showing float and float valve

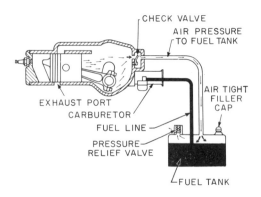

Fig. 4-9 Pressurized fuel systems can be found on some outboard engines.

crankcase to the gas tank. If this method is used, the gas tank must be airtight so that enough air pressure can build up to force the gasoline to flow to the carburetor.

THE CARBURETOR

The carburetor, shown in figures 4-10 and 4-11, must prepare a mixture of gasoline and air in the correct proportions for burning in the combustion chamber. The carburetor must function correctly under all engine speeds, under varying engine loads, in all weather conditions, and at all engine tempera-

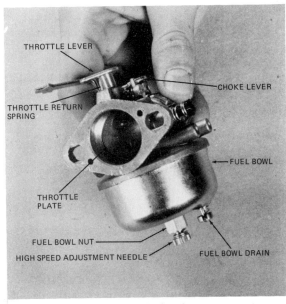

Fig. 4-10 Typical carburetor

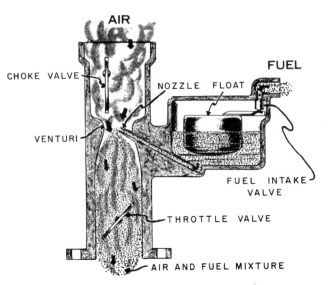

Fig. 4-11 Cross section of a simplified carburetor

tures. To meet all these requirements, carburetors have many built-in parts and systems.

Before studying the carburetor's operation in detail, an examination of the basic carburetor parts and their functions is helpful.

Throttle. The throttle controls the speed of the engine by controlling the amount of fuel mixture that enters the combustion chamber. The more fuel mixture admitted, the faster the engine speed.

Choke. The choke controls the air flow into the carburetor. It is used only for starting the engine. In starting the engine, the operator closes the choke, cutting off most of the engine's air supply. This produces a *rich mixture* (one containing a higher percentage of gasoline) which ignites and burns readily in a cold engine. As soon as the engine starts, the choke is opened.

Needle Valve. The needle valve, figure 4-12, controls the amount of gasoline that is allowed to pass out of the carburetor into the engine. It controls the richness or leanness of the fuel mixture.

Idle Valve. The idle valve also controls the amount of gasoline that is available to the

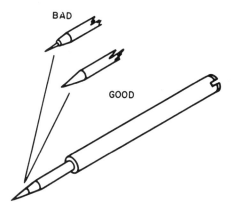

Fig. 4-12 Needle valves

carburetor but it functions only at low speeds or idling. Some carburetors have what is called a *slow-speed needle valve* which performs basically the same job as the idle valve.

Float and Float Bowl. The float and float bowl are found on all carburetors except the suction-fed carburetor and diaphragm carburetors (to be discussed later). The float and float bowl maintain a constant gasoline level in the carburetor.

Venturi. The venturi is a section of the carburetor that is constricted, or has a smaller cross-sectional area for the air to flow through. In the venturi, the gasoline and air are brought together and begin to mix.

Jets. Carburetor jets are small openings through which gasoline passes within the carburtor.

CARBURETOR OPERATION

In the beginning, air flows through the carburetor because there is a partial vacuum or suction in the combustion chamber as the piston travels down the cylinder. One usually thinks of the air being sucked into the engine by the piston action. Atmospheric pressure outside the engine actually pushes the air through the carburetor to equalize the lower pressure that is in the combustion chamber.

Air flows rapidly through the carburetor as the piston moves down, and in the carburetor the air must pass through a constriction called the venturi. See figure 4-13. For the same amount of incoming air to pass through this smaller opening, the air must travel faster, and this it does. Here a principle of physics comes into use: the greater the velocity of air passing through an opening, the lower the static air pressure exerted on the walls of the opening. The venturi thus creates a low-pressure area within the carburetor. See figure 4-14.

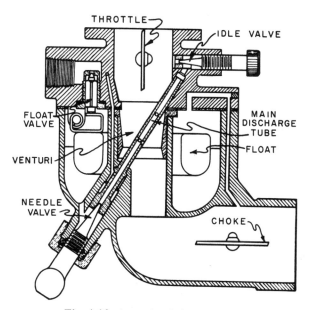

Fig. 4-13 A gravity fed carburetor

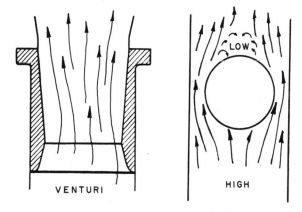

Fig. 4-14 (A) The venturi creates a low pressure in the carburetor. (B) The principle of the air foil also helps create a low pressure.

The principle of the airfoil is also used to gain lower pressure conditions in the venturi. A fuel supply tube or jet is placed in the venturi section. The action of the incoming air causes a high pressure on the front of the jet but a very low pressure on the back of the jet. Gasoline is available in this jet and streams out of the jet because a low-pressure area has been created by the action of the venturi and the airfoil, and because the gasoline is under atmospheric pressure which is greater. Greater pressure pushes the gasoline out of the discharge jet into the airstream. See figure 4-15 for a simplified illustration of carburetor operation.

As gasoline streams into the airflow, it is mixed thoroughly with the air. The best mixture of gasoline and air is fourteen or fifteen parts of air to one part of gasoline, by weight. This air-to-gasoline ratio can be changed for different operating conditions; heavy load and fast acceleration require more gasoline (richer mixture). The needle valve is used to change this ratio; it controls the amount of gasoline that is available to be drawn from the discharge jet.

The throttle or butterfly is mounted on a shaft beyond the venturi section. The operator of the engine controls its setting to regulate the engine speed. When the throttle is wide open the butterfly does not restrict the flow of air; air flows easily through the carburetor and the engine is operating at its top speed. As the operator closes the throttle, the flow of air is restricted. A smaller amount of air can rush through the carburetor, therefore, the air pressures at the venturi section are not as slow and less gasoline streams from the discharge holes. With less fuel mixture in the combustion chamber, the piston is pushed down with less force during combustion; power and speed are reduced.

The ratio of air to fuel remains about the same through the different throttle settings. However, when the throttle is closed and the engine begins to idle, very little air is drawn through the carburetor and the difference between atmospheric pressure and venturi air is slight. Little gasoline is drawn from the discharge jet. In fact, the mixture of gasoline is so lean that a special idling device must be built into the carburetor to provide a richer mixture for idling.

In some carburetors, the main discharge jet is continued past the venturi section to the area of the throttle. It discharges fuel into a small well and jet that are behind the throttle when it is closed. The air pressure behind the throttle is very low. Therefore, gasoline streams from the idle jet readily and mixes with the small amount of air that is coming through the carburetor and a rich fuel mixture is provided for idling. A threaded needle valve called the *idle valve* controls the amount of gasoline that can be drawn from the idle jet. Many carburetors use the air bleed principle to help atomize the gasoline and to make the needle valve easier to adjust. See figure 4-16.

When a cold engine is to be started, an extremely rich mixture of gasoline and air must be provided if the engine is to start easily. The choke, figure 4-17, provides this rich mixture. It is a butterfly placed in the air horn before the venturi section. For

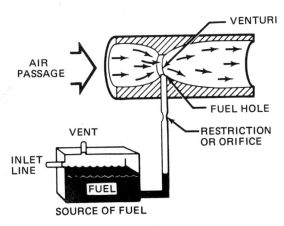

Fig. 4-15 Simplified carburetor operation

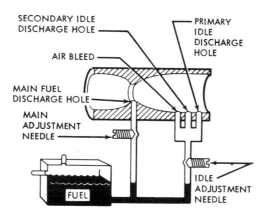

A

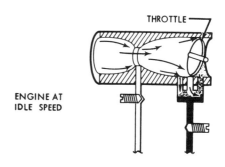

Fig. 4-16 (A) Schematic of basic carburetor parts, (B) engine accelerating, (C) engine at full speed, (D) engine at idle speed, and (E) air bleed principle

starting, the choke is closed, shutting off most of the carburetor's air supply. When the engine is turned over slowly, usually by hand, little air is drawn through the carburetor; but the air pressure within the carburetor is very low and gasoline streams from the main discharge jet, mixing with the air that does get by the choke. As soon as the engine starts, the choke is opened. The fuel mixture provided when the engine is choked is so rich

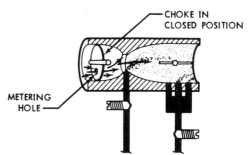

Fig. 4-17 The choke provides a rich mixture for starting.

that all the gasoline may not vaporize with the air and raw liquid gasoline may be drawn into the combustion chamber. Continued operation with the choke closed can cause crankcase dilution—the raw gasoline seeps into the crankcase diluting the lubricating oil.

Therefore, the complete carburetor includes float and float bowl, maintaining a constant reservoir of gasoline in the carburetor; venturi section, producing a low-pressure area; needle valve, controlling the richness or leanness of fuel mixture; main discharge jet, spraying gasoline into the airstream; idle valve, providing a rich mixture for idling conditions; and the choke, producing an extremely rich mixture for easy starting. See figures 4-18 and 4-19.

DIAPHRAGM CARBURETOR

This type of carburetor has come into wide use, especially on chain saws. It is also found on other applications where the engine may be tipped at extreme angles. The diaphragm supplies the carburetor with a constant supply of gasoline.

The diaphragm carburetor may be gravity fed. Crankcase pressure moves the diaphragm and associated linkages to allow gasoline to enter the fuel chamber. The gasoline is available by gravity and the in-and-out motion of the diaphragm meters out fuel in the correct amount.

Diaphragm carburetors are also made with a built-in fuel pump, figures 4-20, page 60,

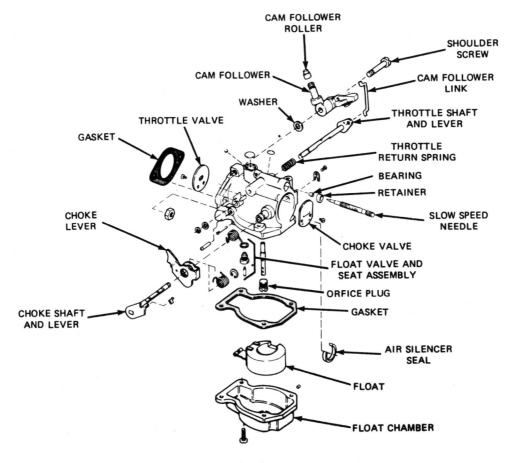

Fig. 4-18 Carburetor components

**FUEL SYSTEM COMPONENTS
AND THEIR FUNCTION**

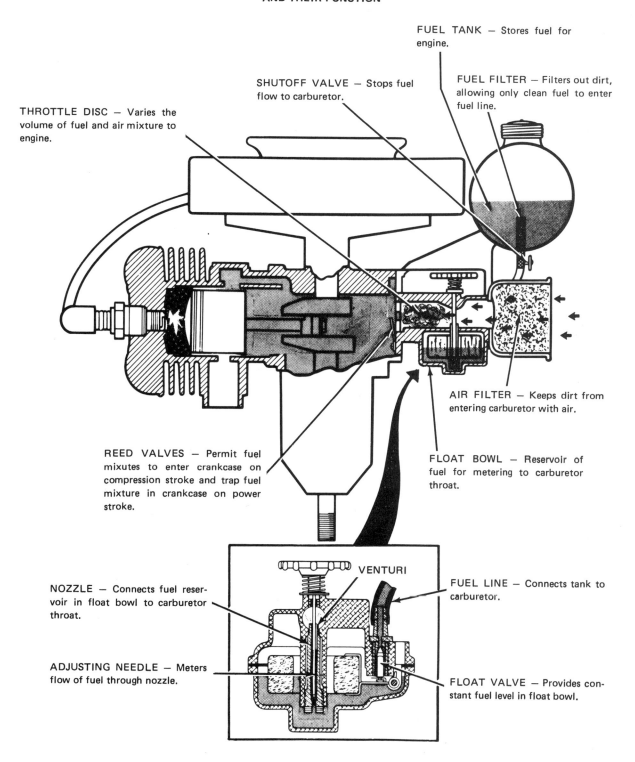

FUEL TANK — Stores fuel for engine.

FUEL FILTER — Filters out dirt, allowing only clean fuel to enter fuel line.

SHUTOFF VALVE — Stops fuel flow to carburetor.

THROTTLE DISC — Varies the volume of fuel and air mixture to engine.

AIR FILTER — Keeps dirt from entering carburetor with air.

REED VALVES — Permit fuel mixutes to enter crankcase on compression stroke and trap fuel mixture in crankcase on power stroke.

FLOAT BOWL — Reservoir of fuel for metering to carburetor throat.

NOZZLE — Connects fuel reservoir in float bowl to carburetor throat.

VENTURI

FUEL LINE — Connects tank to carburetor.

ADJUSTING NEEDLE — Meters flow of fuel through nozzle.

FLOAT VALVE — Provides constant fuel level in float bowl.

Fig. 4-19 Two-stroke cycle carburetor and fuel system

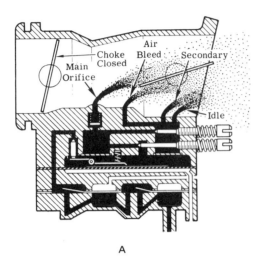

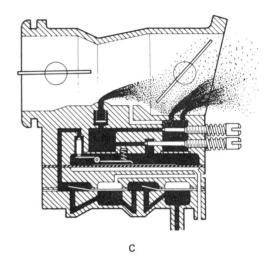

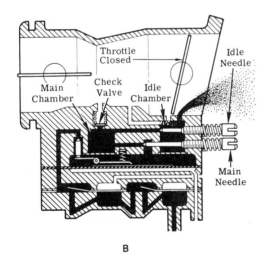

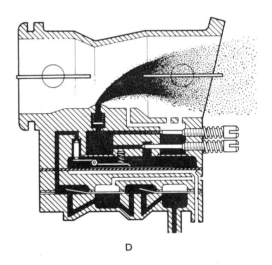

Fig. 4-20 Diaphragm carburetor with built-in fuel pump. (A) choke closed, (B) idle, (C) part throttle, (D) full throttle

and 4-21, page 61. The fuel pump operates on crankcase pressure and its action is much the same as that of the outboard fuel pump previously discussed. Diaphragm carburetors are widely used on small engines.

A common carburetor variation uses both the fuel pump and suction feed principles. The Briggs and Stratton Pulsa-Jet®, figures 4-22 and 4-23, is such a carburetor. At first, it may appear to be an ordinary suction type carburetor, but closer examination shows there are two fuel pipes into the carburetor—one long and one short.

The low pressure created by the intake stroke activates the fuel pump, drawing gasoline through the long pipe and discharging it into a smaller constant level fuel chamber at the top of the tank. This provides a constant level of gasoline regardless of the amount of gasoline in the tank. It also reduces the

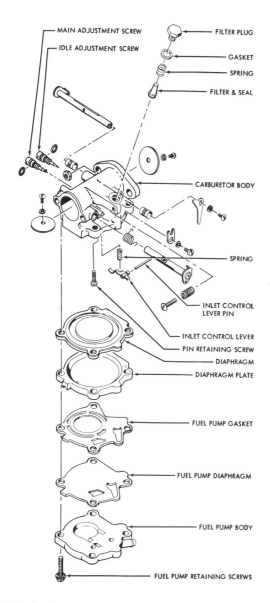

Fig. 4-21 An exploded view of a diaphragm carburetor with a built-in fuel pump

amount of lift required to deliver gasoline into the venturi. This carburetor permits a larger venturi and improves the engine's horsepower rating.

AIR BLEEDING CARBURETORS

There is a tendency for carburetors to supply too rich a fuel mixture at high speeds; the ratio of gasoline to air increases as the velocity of the air passing through the carbu-

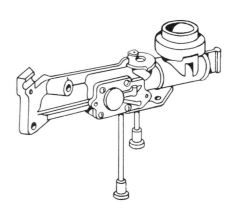

Fig. 4-22 Suction lift carburetor (Pulsa-Jet® by Briggs and Stratton)

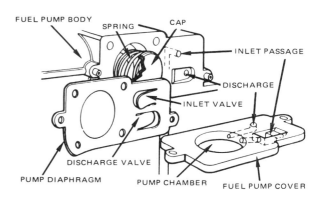

Fig. 4-23 Pulsa-Jet® carburetor pump

retor increases. One method that is commonly used to correct this condition is *air bleeding* (see figure 4-16, view E, page 57). The Zenith® carburetor shown in figure 4-24, page 62, uses the principle of air bleeding.

A small amount of air is introduced into the main discharge well vent to restrict the flow of gasoline from the main discharge jet. As engine speeds are increased, greater amounts of air are brought into the main discharge well vent, placing a greater restriction on the gasoline flow. The additional air overcomes the carburetor's natural tendency to provide too rich a mixture at high speeds. This action maintains the proper ratio of fuel and air between a throttle setting of one-fourth to wide open. The air that enters

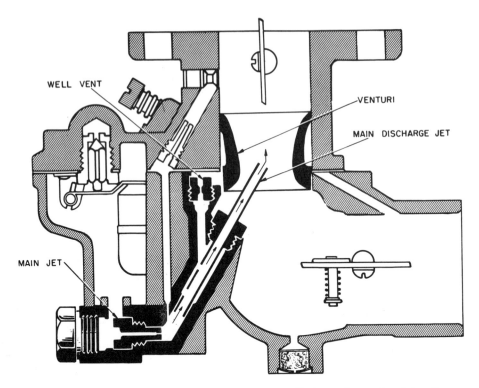

Fig. 4-24 **Air brought into the well vent bleeds into the main discharge jet, maintaining the correct air-fuel ratio throughout throttle range.**

the discharge well vent mixes with the gasoline and is drawn through the main discharge jet into the main airstream.

ACCELERATING PUMP

Another problem inherent in all carburetors is a response lag when the throttle is quickly opened. Air can react very quickly to an increased demand but gasoline lags behind. The result is too lean a mixture and slow acceleration. Carburetors equipped with an accelerating pump provide instant response for rapid acceleration.

The main parts of the accelerating pump, figure 4-25, are the spring, vacuum piston, and fuel cylinder. At idling and low operating speeds, the vacuum piston is drawn to the top of the fuel cylinder by the engine vacuum. This vacuum is strong enough to overcome the force of the spring, holding the piston at the top of its stroke. Now the fuel cylinder is filled with gasoline.

If the throttle is suddenly opened, the engine vacuum drops enough for the piston spring to overcome the force of the vacuum,

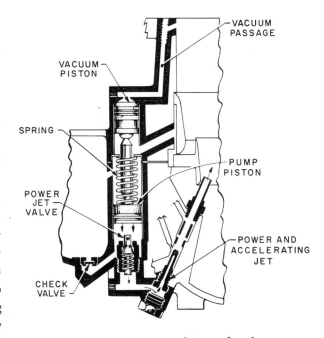

Fig. 4-25 **Cross section of an accelerating pump**

pushing the pump piston down the fuel cylinder. This reserve amount of gasoline in the fuel cylinder is forced into the main discharge jet and on into the carburetor venturi.

CARBURETOR ADJUSTMENTS

The engine manufacturer sets the carburetor adjustments at the factory. These settings cover normal operation. However, after a long period of usage, or under special operating conditions, it may be necessary to adjust the carburetor. In adjusting the carburetor, the main needle valve and the idle valve are both reset to give the desired richness or leanness of fuel mixture. Too lean a mixture can be detected by the engine missing and backfiring. Too rich a mixture can be detected by heavy exhaust and sluggish operation.

Procedure

This is a typical procedure for adjusting a carburetor for maximum power and efficiency.

1. Close the main needle valve and idle valve finger tight. Excessive force can damage the needle valve. Turn clockwise to close.

2. Open the main needle valve one turn. Open the idle valve 3/4 turn. Turn counterclockwise to open.

3. Start the engine, open the choke, and allow the engine to reach operating temperature.

4. Run the engine at operating speed (2/3 to 3/4 of full throttle). Turn the main needle valve in (clockwise) until the engine slows down, indicating too lean a mixture. Note the position of the valve. Turn the needle valve out (counterclockwise) until the engine speeds up and then slows down, indicating too rich a mixture. Note the position of

the valve. Reposition the valve halfway between the rich and lean settings.

5. Close the throttle so the engine runs slightly faster than normal idle speed. Turn the idle valve in (clockwise) until the engine slows down, then turn the idle valve out until the engine speeds up and idles smoothly. Adjust the idle-speed regulating screw (see figure 4-26) to the desired idle speed.

   NOTE: Idle speed is not the slowest speed at which the engine can operate; rather, it is a slow speed that maintains good airflow for cooling and a good take-off spot for even acceleration. A tachometer and the manufacturer's specifications regarding proper idle speed are necessary for the best adjustment.

6. Test the acceleration of the engine by opening the throttle rapidly. If acceleration is sluggish, a slightly richer fuel mixture is usually needed.

CARBURETOR ICING

Carburetor icing, figure 4-27, page 64, can occur when the engine is cold and certain atmospheric conditions are present. If the

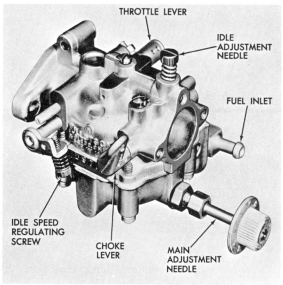

Fig. 4-26 Float-feed type carburetor

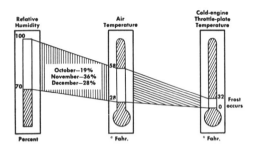

Fig. 4-27   Humidity and temperature conditions which may lead to carburetor icing

temperature is between 28 degrees F. and 58 degrees F., and the relative humidity is above 70 percent, carburetor icing can take place.

Fuel mixture of gasoline and air rushes through the carburetor and the rapid action of evaporating gasoline chills the throttle plate to about 0 degrees F. Moisture in the air condenses and freezes on the throttle plate when the relative humidity is high. See figure 4-28. This formation of ice restricts the air-flow through the carburetor and at low or idle settings the ice can completely block off the airflow, stalling the engine.

When this condition is present, the engine can be restarted but stalls again at low or idle speeds. As soon as the carburetor is warm enough to prevent ice formation, normal operation can take place.

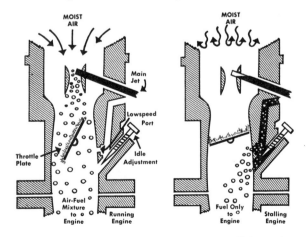

Fig. 4-28   Under carburetor icing conditions, ice forms on the throttle plate, cutting off air to the engine when the throttle closes.

VAPOR LOCK

Vapor lock can occur anywhere along the fuel line, fuel pump, or in the carburetor when temperatures are high enough to vaporize the gasoline. Gasoline vapor in these places cuts off the liquid fuel supply, stalling the engine. If vapor lock occurs, the operator must wait until the carburetor, gas line, and fuel pump cool off and the gasoline vapor returns to liquid before the engine will restart. Vapor lock usually occurs on unseasonably hot days and is more troublesome at high altitudes.

GASOLINE

Most internal combustion engines burn gasoline as their fuel. Gasoline comes from petroleum, also called crude oil. Crude oil is actually a mixture of different hydrocarbons: gasoline, kerosene, heating oil, lubricating oil, and asphalt. These chemicals are all hydrocarbons, but they have characteristics that are quite different. Hydrogen is a light, colorless, odorless gas; carbon is black and solid. Different combinations of carbon and hydrogen give the different characteristics of hydrocarbon products.

The various hydrocarbons are separated by distillation of the crude oil. Crude oil is first heated to a temperature of 700 degrees to 800 degrees F. and then released into a fractionating or bubble tower, figure 4-29. The tower contains 20 to 30 trays through which hydrocarbon vapors can rise from below. When the heated crude oil is released into the tower, most of it flashes into vapor. As the vapors rise, they cool, and each type of hydrocarbon condenses at a different tray level. Heavier hydrocarbons condense first at relatively high temperatures. The lighter hydrocarbons, such as gasoline, condense high in the tower at relatively low temperatures. Gasoline is then further processed to improve its qualities.

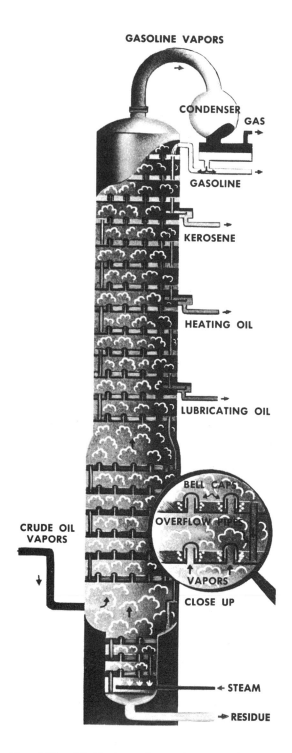

Fig. 4-29 Distillation of petroleum in the bubble tower

Good gasoline must have several characteristics. It must vaporize at low temperatures for good starting. It must be low in gum and sulfur content. It must not deteriorate during storage. It must not knock in the engine. It must have proper vaporizing characteristics for the climate and altitude. It must burn cleanly to reduce air pollution.

The vaporizing ability of gasoline is the key to its success as a fuel. In order to burn inside the engine, the hydrocarbon molecules of the gasoline must be mixed with air since oxygen is also necessary for combustion. To mix the gasoline and air, there must be rapid motion and turbulence to keep the molecules suspended in the air, figure 4-30, page 66. Of course, there is rapid motion and turbulence in the carburetor, and this turbulence continues into the combustion chamber. Combustion chambers are designed to create the maximum turbulence so that gasoline molecules will stay suspended in the air until they are ignited. A thorough mixture ensures smooth and complete burning of gasoline, delivering maximum power.

Although the vaporizing ability of gasoline makes it an excellent fuel, it also presents a great fire and explosion hazard. When vaporized, a gallon of gasoline produces 21 cubic feet of vapor. If this vapor were combined with air in a mixture of 1.4 to 7.6 percent gasoline by volume, it would explode when ignited. Therefore, only one gallon of gasoline properly vaporized would completely fill an average living room with explosive vapor. The safety rules given in unit 2 for the use and storage of gasoline should always be followed.

In engines, the compression of the fuel mixture of gasoline and air results in high combustion pressures and the force necessary to move the piston. This compression may also cause the gasoline to explode *(detonate)* in the engine instead of burning smoothly.

Petroleum fuel — a mixture of many hydrocarbon compounds of different weights.

Like a stack of gravel, consisting of many pebbles basically alike, but of many different sizes—from larger, heavier ones to particles of fine dust.

Extreme turbulence is required to maintain suspension of all particles of fuel vapor — rapid movement through the carburetor and manifold, turbulence in the crankcase and combustion chamber.

Like a whirlwind of gravel on a country road, all particles remain suspended as long as agitation and motion continue.

Heavier particles of the fuel settle or rain out of the fuel vapor with lessening turbulence, to a puddle below.

The very lightest particles (hydrocarbons) remain suspended like the heavier pebbles falling out of the whirlpool and returning to the roadway with lessening of the whirling wind. The very lightest particles remain suspended as dust.

Fig. 4-30 Turbulence is necessary to keep the gasoline molecules suspended in the air.

Detonation is also called knocking, fuel knock, spark knock, carbon knock, and ping. To understand it, one has to think in slow motion. Refer to figure 4-31. The spark plug ignites the fuel mixture, and a flame front moves out from this starting point. As the flame front sweeps across the combustion chamber, heat and pressure build. The unburned portion of the fuel mixture ahead of the flame front is exposed to this heat and pressure. If it self-detonates, two flame fronts are created which race toward each other. The last unburned portion of fuel caught between the two fronts explodes with hammer-like force. Detonation causes a knocking sound in the engine and power loss. Repeated detonation can damage the piston, figure 4-32, page 68.

Preignition causes the same undesirable effects as detonation, including engine damage, figure 4-33, page 68. The cause of preignition, however, is somewhat different. Hot spots, or red-hot carbon deposits, actually ignite the fuel mixture and begin combustion before the spark plug fires.

The antiknock quality of a gasoline, or its ability to burn without knocking, is called its *octane rating*. Gasoline that has no knocking characteristics at all is rated at 100. Usually, tetraethyl lead is added to gasoline to give it the proper antiknock quality. If lead is not used, as in unleaded gasoline, then special aromatic compounds are used to obtain smooth burning.

The higher the compression ratio of the engine, the higher the octane requirements for the engine's fuel. Low-grade gasoline with an octane rating of 70-85 is suitable for compression ratios of 5-7 to 1. Regular grade gasoline with an octane rating of 88-94 is suitable for compression ratios of 7-8.5 to 1. Most small engine manufacturers recommend the use of regular gasoline. High-octane, premuim gasoline does not improve the

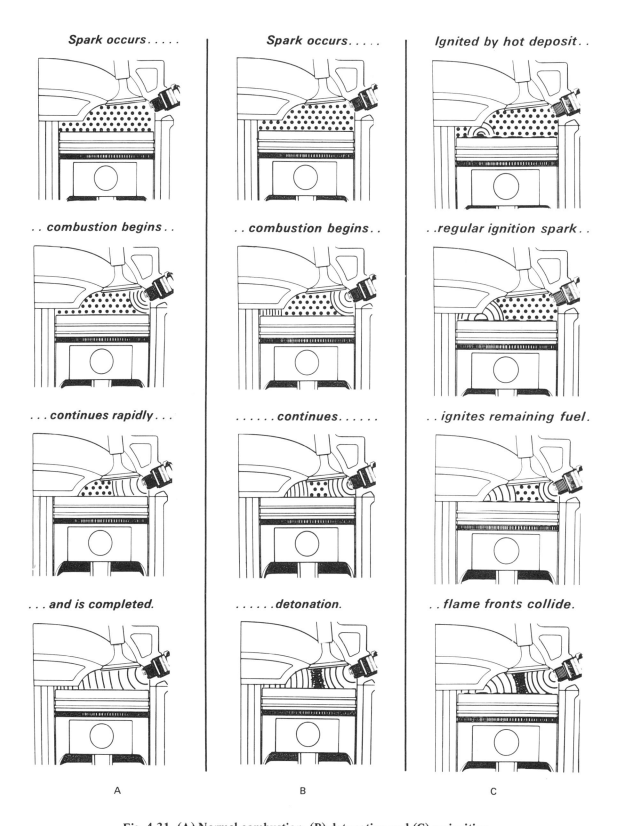

Fig. 4-31 (A) Normal combustion, (B) detonation and (C) preignition

Fig. 4-32 Detonation damage

Fig. 4-33 Preignition damage

performance of small engines. Premium gasoline with an octane rating of about 100 is suitable for compression ratios of 9-10 to 1. Super premium has an octane rating of over 100 and is good for engines with compression ratios of 9.5-10.5 to 1.

ENGINE SPEED GOVERNORS

Speed governors are used to keep engine speed at a constant rate regardless of the load. For instance, a lawn mower is required to cut tall as well as short grass. A governor ensures that the engine operates at the same speed in spite of the varying load conditions. Engines on many other applications need this constant speed feature, too.

Speed governors are also used to keep the engine operating speed below a given, preset rate, so that the engine speed will not surpass this rate. This maximum rate is established and the governor set accordingly by the engineers at the factory. This type of speed governor protects both the engine and the operator from speeds that are dangerously high.

There are two main governor systems used on small gasoline engines: (1) mechanical or centrifugal type and (2) pneumatic or air vane type. Although there are other types of governors, most of the governors used on small engines are included in one of these categories.

MECHANICAL GOVERNORS

The mechanical governor, figures 4-34, 4-35, and 4-36, operates on centrifugal force. With this method, counterweights mounted on a geared shaft, a governor spring, and the associated governor linkages keep the engine speed at the desired RPM. For constant speed operation the action follows this pattern: when the engine is stopped, the mechanical governor's spring pulls the throttle to an open position. The governor spring tends to keep the throttle open, but as the engine speed increases, centrifugal force throws the hinged counterweights further and further from their shaft. This action puts tension on the spring in the other direction, to close the throttle. At governed speed the spring tension is overcome by the counterweights and the throttle will open no further. A balanced position is maintained and the engine assumes its maximum speed.

If, however, a greater load is placed on the engine, it slows down; the hinged counter-

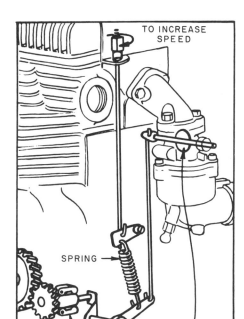
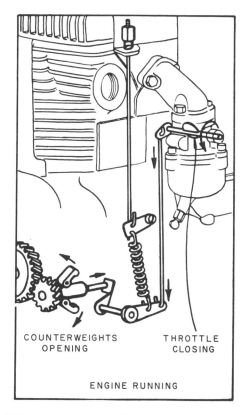

Fig. 4-34 A mechanical governor which operates on centrifugal force

weights swing inward due to lessened centrifugal force; and the governor spring becomes dominant, opening the throttle wider. With a wider throttle opening, the engine speeds up until governed speed is reached again. The action described is fast and smooth; little time is needed for the governor to meet revised load conditions.

The governed speed can be changed somewhat by varying the tension on the governor spring. The more tension there is on the governor spring, the higher the governed speed.

Fig. 4-35 Mechanical governor (centrifugal)

Fig. 4-36 Mechanical governor

Another commonly used centrifugal or mechanical governor is the flyball type. With this governor, round steel balls in a spring loaded raceway move outward as the engine speed increases. The centrifugal force applied by these balls increases until it balances the spring tension holding the throttle open. At this point, the governed speed is reached.

PNEUMATIC OR AIR VANE GOVERNORS

Many engines have a pneumatic or air vane governor, figures 4-37 through 4-39. Again, the governor holds the throttle at the governed speed position, preventing it from opening further. An air vane, which is located near the flywheel blower, controls the speed. As engine speed increases, the flywheel blower pushes more air against the air vane, causing it to change position. At governed speed the air vane position overcomes the governor spring tension and a balance is assumed.

Engines are often designed so that they can be accelerated freely from idle speed up to governed speed. With these engines, the governor spring takes on very little tension at low speeds. The operator has full control. As the throttle is opened further and further, there is more tension on the spring, both from the operator and from the strengthening governor action. At governed speed, the force to close the throttle balances the spring tension to open the throttle.

It should be noted that not all engines have engine speed governors. Engines that

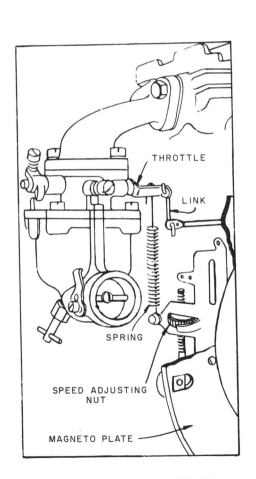
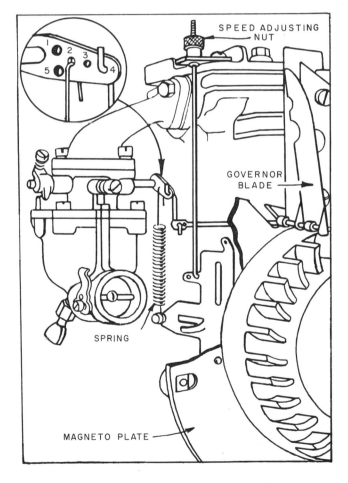

Fig. 4-37 A pneumatic or air vane governor

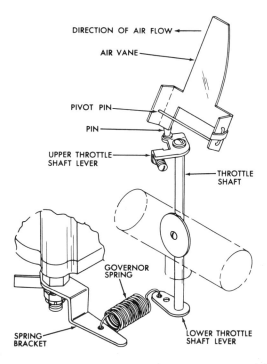

Fig. 4-38 A pneumatic or air vane governor

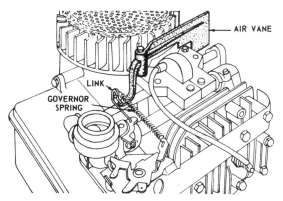

Fig. 4-39 Air vane governor

are always operated under a near constant load do not need the governor. The load of an outboard engine is a good example of a near constant load. Although different loads are placed on the outboard (number of people in boat, position of people in boat, use of trolling mechanism to decrease boat's speed, etc.), the load is constant at full throttle for a given period of use.

REVIEW QUESTIONS

1. Name the basic parts of a typical fuel system.

2. What force is used to operate most outboard fuel pumps?

3. How can a pressurized fuel system be recognized?

4. Briefly describe the task of the carburetor.

5. Name the main parts that make up the carburetor.

6. What causes air to flow through the carburetor?

7. What causes low air pressure in the venturi section of the carburetor?

8. What is the function of the throttle?

9. What is the function of the choke?

10. What is the function of the needle valve?

11. Explain the action of the float in a float-type carburetor.

12. What is the advantage of a diaphragm carburetor?

13. Explain the principle of air bleeding.

14. Explain the operation of the accelerating pump.

15. What is a rich mixture?

16. Why is turbulence important inside the combustion chamber?

17. What two main purposes do engine speed governors serve?

18. What are the two main classifications of engine speed governors?

19. Are engines made that do not have engine speed governors?

20. Can the governed speed be changed? How?

CLASS DEMONSTRATION TOPICS

- Using a fully assembled engine, have the students trace the flow of fuel and also identify the parts of the fuel system.

- Demonstrate, with an engine operating, how to set a carburetor for maximum power.

- Demonstrate, with an operating engine, how a too lean or too rich mixture affects operation and acceleration.

- Demonstrate the action of a carburetor with an atomizer or insect sprayer.

- Disassemble and inspect a fuel pump (automotive or outboard).

- Demonstrate, with an operating engine, how the governor maintains governed speed and does not allow the throttle to open to "full."

- Demonstrate how governed speed can be changed. Stop the engine to readjust the governor.

LABORATORY EXPERIENCE 4-1
BASIC PARTS OF THE CARBURETOR

The carburetor is removed from the engine. It is then disassembled, the basic parts are studied, and it is reassembled. When the carburetor is reassembled, it is reinstalled on the engine. Students are to record their work after each step of the disassembly procedure is completed.

Disassembly Procedure

Instructors may want to supplement or revise specific steps of this procedure since there are many makes of carburetors. The following disassembly procedure is a general guide.

1. Remove the metal shrouding to expose the carburetor, if necessary.
2. Close the fuel shutoff valve at the gas tank or drain the gas tank.
3. Drain the carburetor float bowl, if the drain valve is on the float bowl.
4. Remove the air cleaner. (Use care if it contains oil.)

Part	Disassembly (nuts, bolts, etc.)	Operation performed	Tool used

5. Remove the gas line from the carburetor.

6. Remove the throttle and governor connections.

7. Remove the carburetor from the engine.

8. Disassemble the float bowl and inspect the float chamber.

9. Remove the main or high-speed needle valve (counterclockwise).

10. Remove the idle screw or slow-speed needle valve (counterclockwise).

    CAUTION: Never force needle valves in against their seats—they will be damaged. Do not remove the choke or throttle valves.

REVIEW QUESTIONS

Study the carburetor and answer the following questions.

1. Does the engine have a fuel pump?

2. Is the carburetor suction fed?

3. Is the carburetor gravity fed?

4. Does the carburetor have a float bowl?

5. Does the carburetor have a needle valve?

6. Does the carburetor have an idle valve?

7. Does the carburetor have a choke?

8. Does the carburetor have an air cleaner?

9. What does the carburetor do?

10. What causes air to flow through the carburetor?

11. What does the venturi section do?

12. What is the function of the throttle?

13. What is the function of the choke?

14. Trace the path of fuel and fuel mixture through the engine.

LABORATORY EXPERIENCE 4-2
ADJUSTING THE CARBURETOR FOR MAXIMUM POWER AND EFFICIENCY

The carburetor is adjusted for maximum power and efficiency. These adjustments should be made with the engine operating and with the carburetor and engine at normal operating temperature.

CAUTION: Do not operate the engine without proper ventilation or exhaust system. Also, be certain that no loose clothing can become involved with the moving parts.

Adjustment Procedure

Instructors may want to supplement or revise specific steps of this procedure since there are many makes of carburetors and engines. The following adjustment procedure is a general guide.

1. Close the main needle valve and idle valve finger tight. Excessive force can damage the needle valves. Turn clockwise to close.

2. Open the main needle valve one turn. Open the idle valve 3/4 turn. Turn counter-clockwise to open.

3. Start the engine, open the choke, and allow the engine to reach operating temperature.

4. Run the engine at operating speed (2/3 to 3/4 of full throttle). Turn the main needle valve in (clockwise) until the engine slows down, indicating too lean a mixture. Note the position of the valve. Turn the needle valve out (counterclockwise) until the engine speeds up and then slows down, indicating too rich a mixture. Note the position of the valve. Reposition the valve halfway between the rich and lean setting.

5. Close the throttle so the engine runs slightly faster than normal idle speed. Turn the idle valve in (clockwise) until the engine slows down. Then turn the idle valve back out until the engine speeds up and idles smoothly. Adjust the idle-speed regulating screw to the desired idle speed.

   NOTE: Idle speed is not the slowest speed at which the engine can operate; rather, it is a slow speed that maintains good airflow for cooling and a good takeoff spot for even acceleration. A tachometer and the known proper idle speed are necessary for the best adjustment.

6. Test the acceleration of the engine by opening the throttle rapidly. If acceleration is sluggish, a slightly richer fuel mixture is probably necessary.

REVIEW QUESTIONS

Adjust the carburetor and answer the following questions.

1. What problem does too lean a fuel mixture cause? Too rich?

2. Describe the engine exhaust produced by too rich a fuel mixture.

3. Explain how needle and idle valves can be damaged.

LABORATORY EXPERIENCE 4-3
FUEL PUMPS

The fuel pump is disassembled and inspected, and the diaphragm is observed as it moves in and out. Note how the inlet and discharge valves allow fuel flow in only one direction. Trace the flow of fuel through the pump. Students are to record their work after each step of the disassembly procedure is completed.

Disassembly Procedure

Instructors may want to supplement or revise specific steps of this procedure since students may be working on either an automotive or outboard fuel pump. The following disassembly procedure is a general guide.

1. Remove fuel lines from the tank and to the carburetor.
2. Remove the fuel pump from the engine.
3. Remove machine screws that hold the pump together at the diaphragm.
4. Carefully separate the halves of the pump, exposing the diaphragm.
5. Remove and inspect the inlet and discharge valves (on some pumps this is not possible).

Part	Disassembly (nuts, bolts, etc.)	Operation performed	Tool used

Reassembly Procedure

Reverse the disassembly procedure.

REVIEW QUESTIONS

Study the fuel pump and answer the following questions.

1. Explain what causes the diaphragm to move in and out.

2. Describe the shape of the inlet and discharge valves.

LABORATORY EXPERIENCE 4-4
AIR VANE GOVERNOR (DISASSEMBLE AND INSPECT)

The air vane type of engine governor is examined in order to identify and become familiar with its various parts. Students are to record their work after each step of the disassembly procedure is completed.

Disassembly Procedure

Instructors may want to supplement or revise specific steps of this procedure since there are many makes of engines. The following disassembly procedure is a general guide.

1. Remove all air shrouding that covers the flywheel area, Expose the engine speed governor.

2. Study the air scoops on the flywheel, the governor spring, the air vane, and the associated linkages.

3. Determine how the tension on the governor spring can be changed to increase or decrease governed speed.

4. Remove the governor parts only if directed to do so by the instructor.

Part	Disassembly (nuts, bolts, etc.)	Operation performed	Tool used

Reassembly Procedure

Reverse the disassembly procedure. Use great care with the delicate parts. They must not bind at any point. Free movement is essential for proper operation.

REVIEW QUESTIONS

Study the air vane governor and answer the following questions.

1. What are the main parts of the air vane governor?

2. If the air shrouding were removed from the engine would the governor operation be affected? Why?

3. How can the governor spring tension be changed on the engine?

LABORATORY EXPERIENCE 4-5
ENGINE SPEED GOVERNOR ACTION

The engine speed governor (either mechanical or air vane type) is studied, to familiarize the student with the action of the governor on the engine throttle. The governor is adjusted for the various speed settings. Students are to record their work after each step of the procedure is completed.

Procedure

Either mechanical or air vane governors can be used for this laboratory experience. Use an engine that is on a test stand, not one connected to an implement.

1. With the engine stopped, work the throttle and observe the throttle positions for idle and wide open.

2. Start the engine, bringing the engine to governed speed (many governors do this automatically). Now note the throttle opening. Using a tachometer on the end of the crankshaft, find the engine RPM. Record RPM in a work record box.

3. Stop the engine.

4. Readjust the governor spring tension for increased top speed or readjust the governor spring tension for decreased top speed.

5. Start the engine, bringing the engine to new governed speed. Note the throttle opening. Using a tachometer on the end of the crankshaft find the engine RPM. Record RPM in a work record box.

6. Stop the engine.

Governor Setting	RPM

REVIEW QUESTIONS

Upon completion of the adjustment procedure answer the following questions.

1. Explain the function of the engine speed governor.

2. How was governor spring tension adjusted on the engine?

3. Was the engine speed governor a mechanical or air vane governor?

4. How does the throttle opening in step 5 of the procedure compare with that in step 2?

UNIT 5 LUBRICATION

OBJECTIVES

After completing this unit the student will be able to:

- correctly drain and refill the crankcase with oil.
- discuss and explain the action of ejection pump lubrication.
- discuss the simple splash oil system.

Whenever surfaces move against one another they cause friction and friction results in heat and wear. Lubricating oils have one main job to perform in the engine: to reduce friction. The lubricating oil provides a film that separates the moving metal surfaces and keeps the contact of metal against metal (figure 5-1) to an absolute minimum.

Without a lubricating oil or with insufficient lubrication, the heat of friction builds up rapidly. Engine parts become so hot that they fail. That is when the metal begins to melt, bearing surfaces seize (adhere to other parts), parts warp out of shape, or parts actually break. The common expression is to say the engine *burns up.*

As an example of friction, lay a book on the table and then push the book slowly. Notice the resistance. Friction makes the book difficult to slide. Now place three round pencils between the book and the table top, and push the book. Notice how easily it moves. Friction has been greatly reduced. Oil molecules correspond to the pencils by forming a coating between two moving surfaces. With oil, the metal surfaces roll along on the oil molecules and friction is greatly decreased.

Besides reducing friction and the wear and heat it causes, the lubricating oil serves several other important functions.

Oil Seals Power. The oil film seals power, particularly between the piston and cylinder walls. The great pressures in the combustion chamber cannot pass by the airtight seal which the oil film provides. If this oil film fails, a condition called blow-by exists. Combustion gases push by the film and enter the crankcase. Blow-by not only reduces engine power, but also has a harmful effect on the oil's lubricating quality.

Oil Helps to Dissipate Heat. Oil helps to dissipate heat by providing a good path for heat transfer. Heat conducts readily from inside metal parts through an oil film to outside metal parts that are cooled by air or water. Also, heat is carried away as new oil arrives from the crankcase and the hot oil is washed back to the crankcase.

Oil Keeps the Engine Clean. Oil keeps the engine clean by washing away microscopic

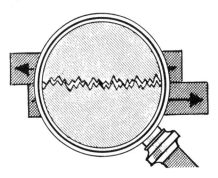

Fig. 5-1 **Exaggerated view of metal surfaces in contact**

pieces of metal that have been worn off moving parts. These tiny pieces settle out in the crankcase or they are trapped in the oil filter, if one is used.

Oil Cushions Bearing Loads. The oil film has a cushioning effect since it is squeezed from between the bearing surfaces relatively slowly. It has a shock absorber action. For example, when the power stroke starts, the hard shock of combustion is transferred ·to the bearing surfaces; the oil film helps to cushion this shock. Figure 5-2 illustrates the oil clearance between the joirnal and the bearing.

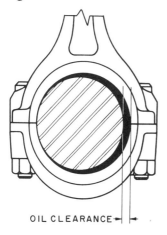

OIL CLEARANCE ►|◄—

Fig. 5-2 Oil clearance between journal and bearing (exaggerated)

Oil Protects Against Rusting. The oil film protects steel parts from rusting. Air, moisture, and corrosive substances cannot reach the metal to oxidize or corrode the surface.

FRICTION BEARINGS

There are three types of friction bearings used in a small gasoline engine: journal, guide, and thrust, figure 5-3. The *journal bearing* is the most familiar. This bearing supports a revolving or oscillating shaft. The connecting rod around the crankshaft and the main bearings are examples. The *guide bearing* reduces the friction of surfaces sliding longitudinally against each other, such as the piston in the cylinder. The *thrust bearing* supports or limits the longitudinal motion of a rotating shaft.

Bearing inserts are commonly used in larger engines or heavy-duty small engines, particularly on the connecting rod and cap and the main bearings. The inserts are precision-made of layers of various metals and alloys. Alloys such as babbit, copper-lead, bronze, aluminum, cadmium, and silver are commonly used. Crankshaft surfaces are, of course, steel. Most small engines, however, do not have bearing inserts, using just the aluminum connecting rod around the steel journal of the crankshaft.

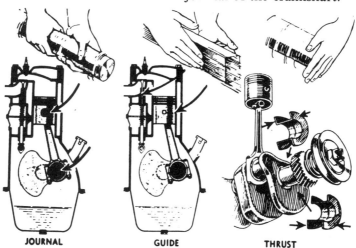

JOURNAL   GUIDE   THRUST

Fig. 5-3 Three major types of friction bearings

ANTIFRICTION BEARINGS

Antifriction bearings, figure 5-4, are also commonly used in engines. They substitute rolling friction for sliding friction. Ball bearings, roller bearings and needle bearings, figure 5-5 are of this type. On many small engines, the main bearings are of the antifriction type.

QUALITY DESIGNATION OF OIL AND SAE NUMBER

Selecting engine lubricating oils can be confusing. One must be aware of different viscosities, different qualities, different refining companies, different additives, and the meaning of many advertising phrases. Generally, lubricating oils used for various

TAPERED ROLLER BEARINGS

NEEDLE BEARINGS

BALL BEARINGS

Fig. 5-4 Types of antifriction bearings

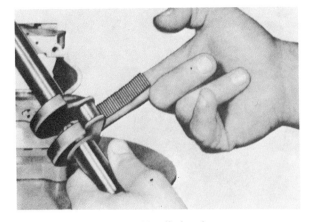

Fig. 5-5 Needle bearings

small gasoline engines are the same ones that are used for automobile engines. However, two-stroke cycle engines almost always use special two-cycle oil.

Quality Designation by the American Petroleum Institute (API)

Figure 5-6 is an example of lubricating instructions given on the decal. The API classifications are usually found on the top of the oil can and refer to the quality of the oil.

Oils for Service SE (Service Extreme). SE oils are suitable for the most severe type of operation beginning with 1972 models and some 1971 automobiles. Extreme conditions of start-stop driving; short trip, cold weather driving; and high speed, long distance, hot weather driving can be handled by this oil. The oil meets the requirements of automobiles that are equipped with emission control devices and engines operating under manufacturers' warranties.

Oils for SD (Service Deluxe). Most small engine manufacturers approve of the use of SD (formerly MS) oil in their engines. These oils provide protection against high and low temperature engine deposits, rust, corrosion, and wear.

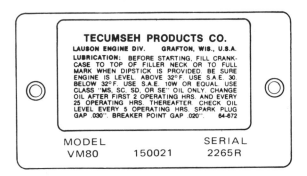

Fig. 5-6 Lubricating instructions on decal

Oils for Service SC. This oil was also formerly classified as suitable for MS. It has much the same characteristics as SD but is not quite as effective as SD oil. SC oil was developed for auto engines of 1964-67, while SD oil was developed for engines of 1968-70 manufacture.

Oils for Service SB (Formerly MM). This oil is recommended for moderate operating conditions such as moderate speeds in warm weather; short-distance, high-speed driving; and alternate long and short trips in cool weather. It is satisfactory for certain older autos but not for new autos under warranty. It is seldom recommended for small engine use.

Oils for Service SA (Formerly ML). No performance requirements are set for this oil. It is straight mineral oil and may be suitable for some light service requirements. SA oil is not recommended by small engine manufacturers.

The new classifications for oils suitable for diesel service are CA, CB, CC, and CD. These replace the old classifications of DG, AM (Supp-1), DM (MIL-L2104B), and DS respectively.

It is wise to buy the best quality oil for an engine. Sacrificing on quality for a few cents savings may be more expensive in the long run due to engine wear.

Viscosity Classification by the Society of Automotive Engineers

The *SAE* (Society of Automotive Engineers) number of an oil indicates its *viscosity,* or thickness. Oils may be very thin (light), or they may be quite thick (heavy). The range is from SAE 5W to SAE 50; the higher the number, the thicker the oil. The owner should consult the engine manufacturer's instruction book for the correct oil. SAE 10W, SAE 20W, SAE 20, and SAE 30 are the most commonly used oil weights. Usually, manufacturers recommend SAE 30 for summer use and either SAE 10W or SAE 20W for cold, subfreezing weather. In winter, thinner oil should be added since oil thickens in cold weather. The *W,* as in 5W, 10W, and 20W, indicates that the oil is designed for service in subfreezing weather.

Multigrade Oils. Multigrade oils, figure 5-7, may also be used in small gasoline engines. These oils span several SAE classifications because they have a very high viscosity index. Typical classifications are SAE 5W-20, or 5W-10W-20; 5W-30 or 5W-10W-20W-30; SAE 10W-30 or 10W-20W-30; 10W-40, and 20W-40. The first number represents the low temperature viscosity of the oil; the last number represents its high temperature viscosity. For example, 10W-30 passes the viscosity test of SAE 10W at low temperatures and the viscosity test of SAE 30 at high temperatures. These oils are also referred to as all-season, all-weather oils, multiviscosity, or multiviscosity grade oils.

Motor Oil Additives

The best oil cannot do its job properly in a modern engine if additives are not blended into the base oil.

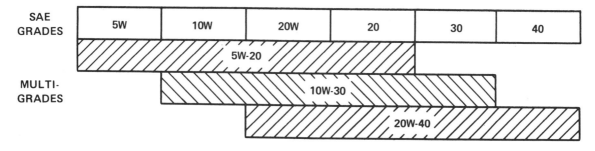

Fig. 5-7 Multigrade motor oils span several SEA (single) grades.

Pour-Point Depressants. Pour-Point depressants keep the oil liquid even at very low temperatures when the wax in oil would otherwise thicken into a buttery consistency making the oil ineffective.

Oxidation and Bearing Corrosion Inhibitors. These additives prevent the rapid oxidation of the oil by excessive heat. Viscous, gummy materials are formed without these inhibitors. Some of these oxidation products attack metals such as lead, cadmium, and silver which are often used in bearings. The inhibiting compounds are gradually used up and, thus, regular oil changes are needed.

Rust and Corrosion Inhibitors. These inhibitors protect against the damage that might be caused by acids and water which are by-products of combustion. Basically, acids are neutralized by alkaline materials, much the same as vinegar can be neutralized with baking soda. Special chemicals surround or capture molecules of water, preventing their contact with the metal. Other chemicals, which have a great attraction for metal, form an unbroken film on the metal parts. These inhibitors are also used up in time.

Detergent/Dispersant Additives. This type of additive prevents the formation of sludge and varnish. Detergents work much the same as household detergents in that they have the ability to disperse and suspend combustion contaminants in the oil but do not affect the lubricating quality of the oil. When the oil is changed, all the contaminants are discarded with the oil so that the engine is kept clean. Larger particles of foreign matter in the oil either settle out in the engine base or are trapped in the oil filter if one is used.

Foam Inhibitors. Foam inhibitors are present in all high-quality motor oils to prevent the oil from being whipped into a froth or foam. The action in the crankcase tends to bring air into the oil, and foaming oil is not an effective lubricant. Foam inhibitors called silicones have the ability to break down the tiny air bubbles and cause the foam to collapse.

LUBRICATION OF FOUR-CYCLE ENGINES

Since all moving parts of the engine must be lubricated to avoid engine failure, a constant supply of oil must be provided. Each engine, therefore, carries its own reservoir of oil in its crankcase where the main engine parts are located.

Basically, the oil is either pumped to or splashed on the parts and bearing surfaces that need lubrication. There are several lubrication systems used on small engines and the systems discussed below are among the most common.

* Simple Splash
* Constant-Level Splash
* Ejection Pump
* Barrel-type Pump
* Full-Pressure Lubrication

Splash System

The *splash system* is perhaps the simplest system for lubrication. It consists of a splasher or dipper that is fastened to the connecting rod cap. There are several different sizes and shapes of dippers, figure 5-8. The dipper may

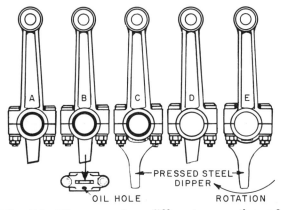

Fig. 5-8 Dippers used on different connecting rods

be bolted on with the connecting rod cap or may be cast as a part of the connecting rod cap. Each time the piston nears the bottom of its stroke the dipper splashes into the oil reservoir in the crankcase, splashing oil onto all parts inside the crankcase. Since the engine is operating at 2000 to 3000 RPM, the parts are literally drenched by millions of oil droplets. Some engines use an oil slinger, figure 5-9, that is driven by the camshaft. The slinger performs a similar function to the dipper.

Constant-Level Splash System

The *constant-level splash system*, figure 5-10, has three refinements over the simple

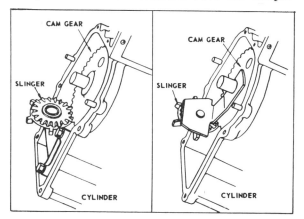

Fig. 5-9 Oil slinger

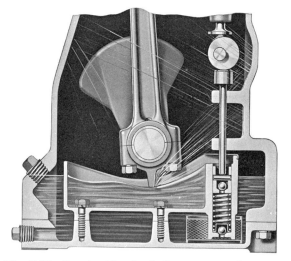

Fig. 5-10 Constant-level splash system used on some engines

splash system: a pump, a splash trough, and a strainer. With this system a cam-operated pump brings oil from the bottom of the crankcase into a splash trough. Again the splasher on the connecting rod dips into the oil, splashing it on all parts inside the crankcase. The strainer prevents any large pieces of foreign matter from recirculating through the system. The pump maintains a constant oil supply in the trough regardless of the oil level in the crankcase.

Ejection Pump

Ejection pumps, figure 5-11, of various types are found on many small engines. With this method, a cam-operated pump draws oil from the bottom of the crankcase and sprays or squirts it onto the connecting rod. Some of the oil enters the connecting rod bearing through small holes, while the remainder is deflected onto the other parts within the crankcase.

The parts of the pump are the base, screen, check valve, spring, and plunger. The spring tends to push the plunger up but the plunger is driven down every revolution by the cam. A check valve allows the chamber to be filled with oil when the plunger goes up but when the cam pushes the plunger down,

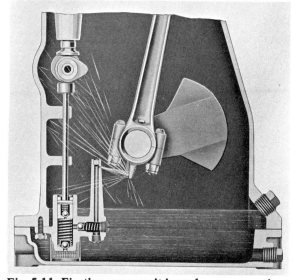

Fig. 5-11 Ejection pump as it is used on some engines

the check valve closes, and the trapped oil is squirted or sprayed from the pump.

The *barrel-type pump,* figures 5-12 and 5-13, is also driven by an eccentric on the camshaft. The camshaft is hollow and extends to the pump of the vertical crankshaft. As the pump plunger is pulled out on intake, an intake port in the camshaft lines up, allowing the pump body to fill. When the plunger is forced into the pump body on discharge, the discharge ports in the camshaft line up, allowing the oil to be forced to the main bearing and to the crankshaft connecting rod journal. Small

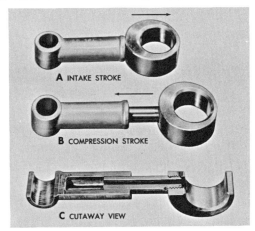

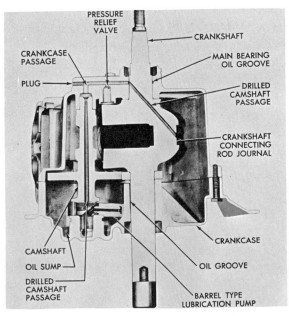

Fig. 5-13 Lubrication oil flow of barrel-type lubrication system

drilled passages are used to channel the oil. Oil is also splashed onto other crankcase parts.

Full-pressure lubrication, figure 5-14, is found on many engines, especially the larger of the small engines and on automobile engines. Lubricating oil is pumped to all main, connecting, and camshaft bearings through small passages drilled in these engine parts. Oil is also delivered to tappets, timing gears, etc. under pressure. The pump used is usually a positive displacement gear type. It is also common to use a splash system in conjunction with full-pressure systems.

LUBRICATING CYLINDER WALLS

Oil is splashed or sprayed onto cylinder walls and the piston rings spread the oil evenly for proper lubrication. Piston rings must function properly to avoid excessive oil consumption. The rings must put an even pressure on the cylinder walls and give a good seal. If piston ring grooves are too large, the piston rings begin a pumping action as the piston moves up and down. As a result, excess oil is brought into the combustion chamber. Worn cylinder walls, worn pistons, and worn rings can all add to high oil con-

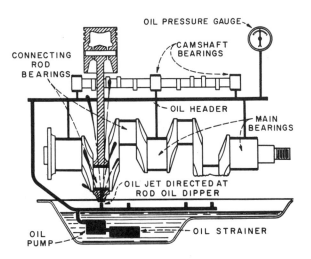

Fig. 5-14 Full-pressure lubrication

sumption as well as loss of compression and eventual loss of power.

BLOW-BY

Blow-by, the escape of combustion gases from the combustion chamber to the crankcase, occurs when piston rings are worn or too loose in their grooves. Carbon and soot from burned fuel are forced into the crankcase by the rings. Much of the carbon is deposited around the rings, hindering their operation further.

Another damaging result of worn rings can be crankcase dilution. If raw, unburned gasoline collects in the combustion chamber it can leak by the piston rings and into the crankcase. This dilutes the crankcase oil, and, thereby, reduces the oil's lubricating properties.

CRANKCASE BREATHERS

Four-stroke cycle engines do not have completely airtight, oiltight crankcases. Engine crankcases must breathe. Without the ability to breathe, pressures build up in the crankcase and cause oil seals to rupture or allow contaminants to remain in the crankcase. Pressure buildup may be caused by the expansion of the air as the engine heats up, by the action of the piston coming down the cylinder, and by the blow-by of combustion gases along cylinder walls.

Most single-cylinder engines use breathers, figure 5-15, that allow air to leave but not reenter a reed-type check valve or a ball-type check valve. These breathers place the crankcase under a slight vacuum and are called closed breathers.

Open breathers allow the engine to breathe in and out freely and are usually equipped with air filters. They are located where the splashing of oil is not a problem. Open breathers are often incorporated with

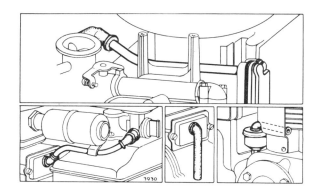
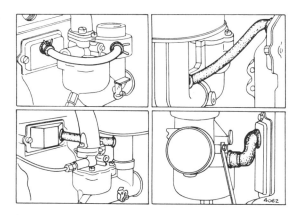

Fig. 5-15 Breather assemblies

the valve access cover. If an engine with this type of breather is tipped on its side, oil may run out through it. Some breathers are vented to the atmosphere while others are vented back through the carburetor.

LUBRICATION OF TWO-CYCLE ENGINES

Lubrication of the two-cycle engine is quite different from the four-cycle engine. Since the fuel mixture must travel through the crankcase, a reservoir of oil cannot be stored there. The lubricating oil is mixed with gasoline and then put into the gas tank. The lubricating oil for all crankcase parts enters the crankcase as a part of the fuel mixture. Millions of tiny oil droplets suspended in the mixture of gasoline and air settle onto the moving parts in the crankcase, providing lubrication. Oil droplets, being relatively large and heavy, quickly drop out of suspension. Of course, much oil is carried on into the combustion chamber where it is burned along with the gasoline and air.

On a two-cycle engine, oil must be mixed with the gasoline. The engine is not operated on gasoline alone. If it were, the heat of friction would burn up the engine in a short time. Some two-cycle engine manufacturers are developing and marketing engines with oil metering devices that eliminate the need to premix the oil and gasoline. Mixing is done automatically in the correct proportions.

In preparing the mixture of gasoline and oil for most two-cycle engines, observe the following rules:

- Mix a good grade of regular gasoline and oil in a separate container, figure 5-16. Do not mix in the gas tank unless it is a remote tank such as is found on many outboards.

- Pour the oil into the gasoline to ensure good mixing, and shake the container well. If poorly mixed, the oil settles to the bottom of the tank, causing hard starting.

- Strain the fuel with a fine mesh strainer when pouring it into the tank to prevent any moisture from entering the tank.

- Use the oil that is specified for the particular two-cycle engine. Most manufacturers specify their own brand. In an emergency, other oils may be used, usually SAE 30SB (formerly MM) or SAE 30SD (formerly MS) nondetergent oil. The manufacturer's brand, however, provides the best lubrication with a minimum of deposit formation.

- Mix gasoline and oil in the proportions recommended by the engine manufacturer. One common proportion is three-fourths pint of oil to one gallon of gasoline when breaking in a new engine, and

Fig. 5-16 Thorough mixing of oil and gasoline is necessary for two-cycle engines.

one-half pint of oil to one gallon of gasoline for normal use.

NOTE: Mixing proportions vary widely; check the engine's instruction book.

ADDITIONAL LUBRICATION POINTS

Whether the engine has two-cycle or four-cycle lubrication, there may be other lubrication to consider besides the crankcase area. On an outboard engine do not neglect the lower unit, figure 5-17, which needs a special gear lubricant at several points. If an engine is powering an implement or other machinery, there may be transmissions, gear boxes, chains, axles, wheels, shafts, and linkages, for example, that need periodic lubrication. Lubricate these additional parts with the oil or lubricant recommended by the manufacturer.

OIL CHANGES

The crankcase oil should be changed periodically. The exact number of engine operating hours between oil changes varies depending on the manufacturer. It may be

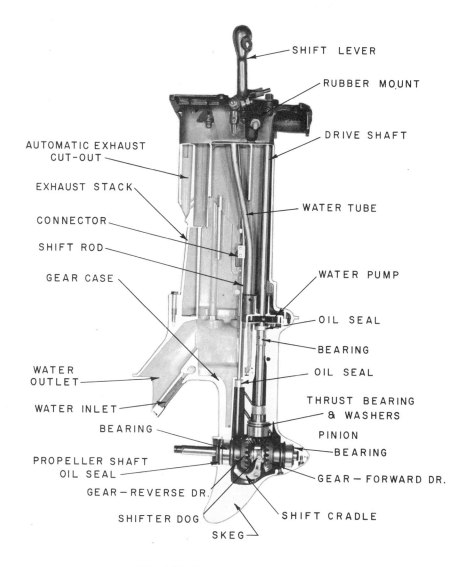

SHIFT LEVER

RUBBER MOUNT

DRIVE SHAFT

AUTOMATIC EXHAUST
CUT-OUT

EXHAUST STACK

CONNECTOR

SHIFT ROD

GEAR CASE

WATER TUBE

WATER PUMP

OIL SEAL

BEARING

OIL SEAL

THRUST BEARING
& WASHERS

PINION

WATER
OUTLET

WATER INLET

BEARING

BEARING

GEAR — FORWARD DR.

PROPELLER SHAFT
OIL SEAL

GEAR — REVERSE DR.

SHIFTER DOG

SKEG

SHIFT CRADLE

Fig. 5-17 Gear case—lower unit assembly

as short a time as every 20 hours or as long a time as every 100 hours. (Every 25 hours is most common.) To ensure minimum engine wear and maximum engine life, follow the manufacturer's lubrication suggestions.

It is best to drain the oil when the engine is hot since more dirt and slightly more oil can be removed. Dirty oil should be replaced because it cannot give proper, high-quality lubrication.

REVIEW QUESTIONS

1. What is the main job of a lubricant?

2. Explain how the lubricating oil also:

a. Seals power

b. Helps to dissipate heat

c. Keeps the engine clean

d. Cushions bearing loads

e. Protects against rusting

3. What are the three types of friction bearings?

4. What is an antifriction bearing?

5. What does an oil's SAE number refer to?

6. What is a multigrade oil?

7. Explain how a detergent oil works.

8. What are the quality designations of oil?

9. List five common four-cycle engine lubrication systems.

10. How are cylinder walls lubricated?

11. Explain blow-by.

12. Briefly explain how two-cycle lubrication is accomplished.

13. What is one common proportion of oil to gasoline?

14. What kind of oil should be used for two-cycle engines?

CLASS DISCUSSION TOPICS

- Discuss how friction causes heat and wear.
- Discuss the jobs a lubricant performs.
- Discuss the types of bearings used on a demonstration engine.
- Discuss the good features of the various lubrication systems; also discuss the system's weak points, if any.

CLASS DEMONSTRATION TOPICS

- Illustrate the theory of lubrication using round pencils and a book.
- Demonstrate oil viscosity by showing and pouring several different weights of oil.
- Remove and inspect oil control rings.
- Remove an ejection pump from an engine and demonstrate the pump's operation.
- Remove and inspect the dipper on a splash lubrication system.

LABORATORY EXPERIENCE 5-1
DRAIN OIL AND REFILL CRANKCASE

The crankcase oil is drained out and replaced with the correct type of oil. Take care to prevent a mess; a good worker is not sloppy. Students are to record their work after each step of the procedure is completed.

Procedure

Instructors may want to supplement or revise specific steps of this procedure since there are many types of engines. The following procedure is a general guide.

1. Remove the add-oil plug.

2. Loosen and carefully remove the drain plug. Do not drop the plug when it comes out of the engine. Be sure to have a container directly under the drain hole.

3. Allow the oil to drain, then tip the engine slightly to get the last bit of oil from the engine.

4. Replace the drain plug.

5. Refill the crankcase with clean oil of the correct quality and viscosity.

6. Check the oil level with the dipstick if the engine has one. On engines not having a dipstick, fill the crankcase until the oil can be seen at the hole or in the fill pipe.

7. Replace add-oil plug.

Part	Disassembly (nuts, bolts, etc.)	Operation performed	Tool used

REVIEW QUESTIONS

Complete the work on page 99, then answer the following questions.

1. On most engines, how often should lubricating oil be changed?

2. What quality oil is generally used?

3. What viscosity (SAE No.) oil is often used in the summer? Winter?

4. Why is it a good idea to drain the oil just after the engine has been operated and it is hot?

LABORATORY EXPERIENCE 5-2
EJECTION OIL PUMP

Note: Any engine with a camshaft driven pump that sprays or squirts oil onto the parts or pumps oil into the splash trough is suitable for this laboratory experience.

The engine is disassembled, exposing the pump, and the pump is then removed from the engine. The construction of the pump is studied and the operation is observed, while the pump is slowly being worked by hand. Students are to record their work after each step of the disassembly procedure is completed.

Disassembly Procedure (Ejection Pump)

Instructors may want to supplement or revise specific steps of this procedure since there are many makes of engines. The following procedure is a general guide.

1. Remove the air shrouding if necessary.
2. Drain the fuel system and oil bath air cleaner.
3. Drain the oil from the crankcase.

Part	Disassembly (nuts, bolts, etc.)	Operation performed	Tool used

4. Remove the crankcase from the base or sump. Note: there is a gasket between these sections.

5. Looking up into the bottom of the crankcase, locate the crankshaft, camshaft, and oil pump.

6. Remove the oil pump from the engine.

7. Disassemble the pump and study its main parts.

8. Reassemble the pump.

9. Study pump operation by:

   a. Pouring the oil into the engine base.

   b. Placing the oil pump base in the oil supply.

   c. Using a finger on the top of the plunger as a substitute for cam action, to work the pump.

   CAUTION: Have a shield to stop the oil if it comes out of the pump with too much force.

Reassembly Procedure

Reverse the disassembly procedure. A new gasket may be needed between the base and crankcase.

REVIEW QUESTIONS

Study the ejection oil pump and its operation and then answer the following questions.

1. What are the main parts of the pump?

2. Is the oil discharge directed at any particular part within the crankcase or just the crankcase in general?

3. Why is a gasket necessary between the base and crankcase?

4. If the engine is operating at 3600 RPM, how many pump strokes are there each minute?

5. Can a drop in the level of crankcase oil affect the pump's output? Explain.

LABORATORY EXPERIENCE 5-3
SIMPLE SPLASH OIL SYSTEM (HORIZONTAL CRANKSHAFT)

The engine is disassembled, exposing the oil dipper. The crankshaft is then slowly revolved and the path of the dipper is studied. The dipper is removed and its construction is studied. Students are to record their work after each step of the disassembly procedure is completed.

Disassembly Procedure

Instructors may want to supplement or revise specific steps of this procedure since there are many makes of engines. The following procedure is a general guide.

1. Remove the air shrouding if necessary.
2. Drain the fuel system and oil bath air cleaner.
3. Drain the oil from the crankcase.
4. Remove the crankcase from the base or sump (on some engines remove the cover assembly from the crankcase). Note: there is a gasket between these sections.
5. Locate the oil dipper, slowly revolve the crankshaft, and observe the path of the dipper.
6. Remove the connecting rod cap and dipper assembly. Study the part.

Part	Disassembly (nuts, bolts, etc.)	Operation performed	Tool used

Reassembly Procedure

Reverse the disassembly procedure. Use care to replace the dipper and connecting rod cap exactly as they came off. A new gasket may be needed between the base and crankcase.

REVIEW QUESTIONS

Study the splash oil system and then answer the following questions.

1. Is the dipper on the engine case part of the connecting rod cap or is it a separate part?

2. If a one-cylinder engine is operating at 4000 RPM, how many times does the dipper splash into the oil reservoir per minute?

3. Can a drop in the level of crankcase oil affect the lubrication within the crankcase? Explain.

UNIT 6 COOLING SYSTEMS

OBJECTIVES

After completing this unit the student will be able to:

- discuss the construction of the water jacket and trace the path of cooling water through the engine.

- explain the construction and operation of the water pump and trace the path of water through the pump.

- list the basic parts of the air cooling system and trace the path of cooling air through the engine.

In internal combustion engines, the temperature of combustion often reaches over 4000 degrees Fahrenheit, a temperature well beyond the melting point of the engine parts. This intense heat cannot be allowed to build up. A carefully engineered cooling system is, therefore, a part of every engine. A cooling system must maintain a good engine operating temperature, figure 6-1, without allowing destructive heat to build up and cause engine part failure.

The cooling system does not, of course, have to dispose of all the heat produced by combustion. A good portion of the heat

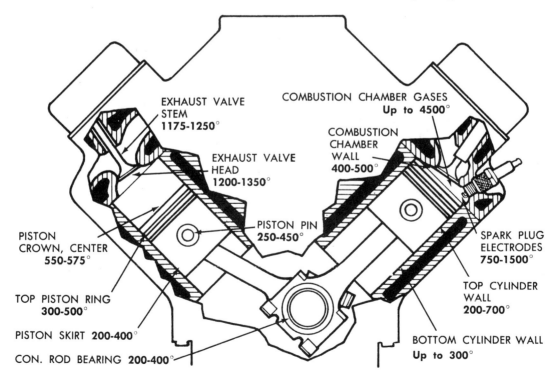

Fig. 6-1 Approximate engine operating temperatures

energy is converted into mechanical energy by the engine; benefiting engine efficiency. Some heat is lost in the form of hot exhaust gases. The cooling system, however, must dissipate about one-third of the heat energy caused by combustion.

Engines are either air-cooled or water-cooled; both systems are in common use. Generally, air-cooled engines are used to power machinery, lawn mowers, garden tractors, chain saws, etc. The air-cooled system is usually lighter in weight and simpler; hence, its popularity for portable equipment.

The water-cooled engine is often used for permanent installation or stationary power plants. Most automobile engines and outboard motors are water-cooled.

AIR-COOLING SYSTEM

The air-cooling system consists of heat radiating fins, flywheel blower, and shrouds for channeling the air. The path of airflow can be seen in figure 6-2.

Heat radiating fins are located on the cylinder head and cylinder because the greatest concentration of heat is in this area. The fins increase the heat radiating surface of these parts allowing the heat to be carried away more quickly. See figure 6-3.

The flywheel blower consists of air vanes cast as a part of the flywheel. As the flywheel revolves, these vanes blow cool air across the fins, carrying away the heated air and replacing it with cool air.

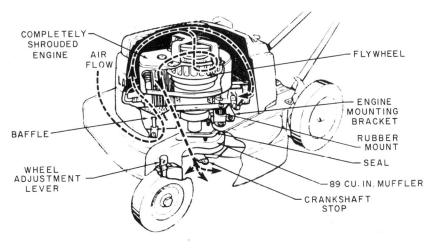

Fig. 6-2 The path of airflow on certain lawn mower engines

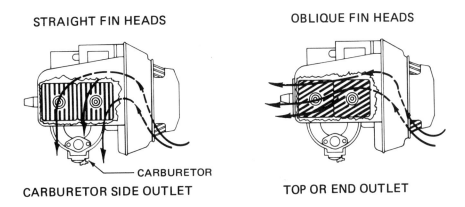

Fig. 6-3 Airflow across the cylinder head fins

The shrouds, figure 6-4, direct the path of the cool air to the areas that demand cooling. Shrouds must be in place if the cooling system is to operate at its maximum efficiency.

CARE OF THE AIR-COOLING SYSTEM

The air-cooling system is almost trouble free. Several points, however, should be considered. Heat radiating fins are thin and often fragile, especially on aluminum engines. If, through carelessness, they are broken off, a part of the cooling system is gone. Besides losing some cooling capacity, hot spots can develop, warping the damaged area. Also, it is easy for dust, dirt, grass, and oil to accumulate between the fins. Any buildup of foreign matter reduces the cooling system's efficiency. All parts of the system should be kept clean, especially the radiating fins.

The flywheel vanes should not be chipped or broken. Besides reducing the cooling capacity of the engine, such damage can destroy the balance of the flywheel. An un-balanced flywheel causes vibration and an excessive amount of wear on engine parts.

WATER-COOLING SYSTEM

Many small engines are water-cooled as are most larger engines. Such engines have an enclosed water jacket around the cylinder walls and cylinder head. Water jackets are relatively trouble free; however, damaging deposits can build up over a period of time. Salt corrosion, scale, lime, silt, etc., can restrict the flow of water and heat transfer. Water jackets are found on all outboard motors and on the majority of automobile engines. Cool water is circulated through this jacket, picking up the heat and carrying it away.

Figure 6-5 shows a cross section of water passage in the head and block. Water cooling systems often found on stationary small gas engines or automobile engines include the following basic parts: radiator, fan thermostat, water pump, hoses, and water jacket.

The water pump circulates the water throughout the entire cooling system. The hot water from the combustion area is carried from the engine proper to the radiator. In

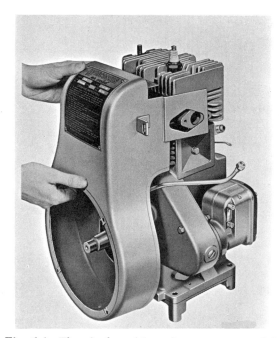

Fig. 6-4 The air shroud is an important part of the air-cooled engine.

Fig. 6-5 Cross section showing water passages in head and block

the radiator, many small tubes and radiating fins dissipate the heat into the atmosphere. A fan blows cooling air over the radiating fins. From the bottom of the radiator the cool water is returned to the engine.

Engines are designed to operate with a water temperature of between 160 degrees and 180 degrees Fahrenheit. To maintain the correct water temperature a thermostat is used in the cooling system. When the temperature is below the thermostat setting, the thermostat remains closed (figure 6-6A) and the cooling water circulates only through the engine. However, as the heat builds up to the thermostat setting, the thermostat opens (figure 6-6B) and the cooling water moves throughout the entire system.

WATER COOLING THE OUTBOARD ENGINE

Cooling the outboard engine with water is a simpler process because there is an inexhaustible supply of cool water present where the engine operates. Outboards pump water from the source through the engine's water jacket and then discharge the water back into the source. A cutaway of an outboard motor is shown in figure 6-7.

The water pump on outboards is located in the lower unit. It is driven by the main driveshaft or the propeller shaft. The cool

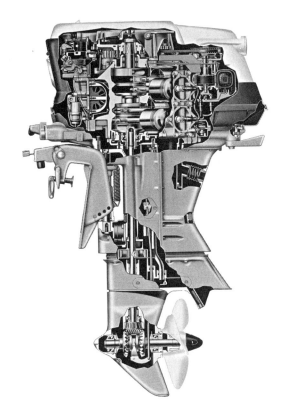

Fig. 6-7 Cutaway of an outboard motor

water is pumped up copper-tube passages to the water jacket. After the cooling water picks up heat, it is discharged into the exhaust area of the lower unit and out of the engine. Several types of water pumps are used on outboard motors: plunger type pumps, eccentric rotor pumps (figure 6-8), impeller pumps (figure 6-9, page 110), and others.

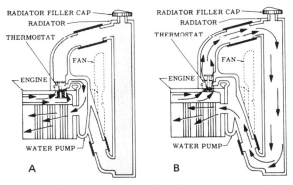

Fig. 6-6 (A) Thermostat closed: water recirculated through engine only. (B) Thermostat open: water circulated through both engine and radiator

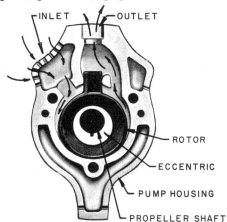

Fig. 6-8 An eccentric rotor water pump with the cover removed to show the eccentric and rotor

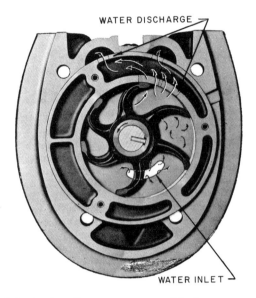

Fig. 6-9 An impeller water pump with the cover removed to show the impeller

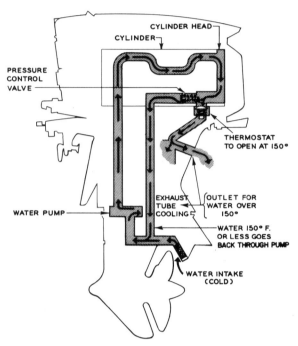

Fig. 6-10 A thermostatically controlled cooling system used on outboard engines

Many outboard motors, especially large horsepower models, are equipped with a thermostatically controlled cooling system, figure 6-10. The temperature of the water circulating through the water jacket is maintained at about 150 degrees Fahrenheit.

REVIEW QUESTIONS

1. How high can the temperature of combustion reach?

2. What are the melting points of aluminum and iron?

3. Does the cooling system remove all the heat of combustion? Explain.

4. Why are air-cooling systems often used for portable equipment?

5. What are the main parts of the air-cooling system?

6. Why are cylinders and cylinder heads equipped with fins instead of being cast with a smooth surface?

7. What are the main parts of the water-cooling system?

8. What is the function of the thermostat?

9. What are several types of water pumps used on outboard motors?

CLASS DISCUSSION TOPICS
- Discuss how heat can damage engine parts.
- Discuss the path of heat flow from the inside of the engine to the outside.
- Discuss the advantages of the air-cooled engine.
- Discuss the advantages of the water-cooled engine.
- Discuss why it is best for the engine to operate at a constant temperature.

CLASS DEMONSTRATION TOPICS
- Trace the airflow through an air-cooled engine.
- Trace the water flow through an outboard engine.
- Disassemble various types of water pumps; show their operation and construction.
- Show how the air-cooling system can be damaged.

LABORATORY EXPERIENCE 6-1
WATER JACKET OF OUTBOARD MOTORS

The water jacket of an outboard engine is disassembled and inspected. The path of the cooling water within the jacket is determined. Use a small horsepower outboard engine if possible. Students are to record their work after each step of the disassembly procedure is completed.

Disassembly Procedure

Instructors may want to supplement or revise specific steps of this procedure since there are many makes of outboards. The following procedure is a general guide.

1. Remove the shrouds.
2. Remove the spark plugs.
3. Remove the machine screws in the top of the water jacket.
4. Remove the top of the water jacket. (Note: the gasket between the engine block and the top of the water jacket is usually glued on and may be difficult to remove.)
5. Examine the construction of the water jacket. Note how deep it is and trace the path of the water flow.

Part	Disassembly (nuts, bolts, etc.)	Operation performed	Tool used

Reassembly Procedure

Reverse the disassembly procedure. A new gasket has to be installed if the engine is to be operated. All metal gasket surfaces must be absolutely clean before a new gasket is installed.

REVIEW QUESTIONS

Study the water jacket and then answer the following questions.

1. What can harm or damage the water jacket?

2. What would happen if an outboard engine were operated out of water?

3. What are the basic parts of the water-cooled engine?

LABORATORY EXPERIENCE 6-2
WATER PUMP OF OUTBOARD MOTORS

The water pump of an outboard engine is disassembled and inspected. There is such a wide variance in the location and disassembly procedure that it is not practical to include a general disassembly procedure in this assignment. Therefore, the instructor may advise which parts to remove and how to do the work. Students are to record their work after each step of the disassembly is completed.

Part	Disassembly (nuts, bolts, etc.)	Operation performed	Tool used

REVIEW QUESTIONS

Study the water pump and answer the following questions.

1. What type of water pump does the engine have?

2. What drives the water pump?

3. Describe the location of the water intake.

4. Is the engine's cooling system thermostatically controlled?

5. Why is it important to check that the engine is pumping water after starting it?

LABORATORY EXPERIENCE 6-3
AIR COOKING SYSTEM

The basic parts of the air-cooling system are disassembled and inspected. The parts are inspected for damage and cleanliness. Students are to record their work after each step of the disassembly procedure is completed.

If proper ventilation and facilities are available for running the engine, demonstrate the airflow through the system before the disassembly and inspection portion of this assignment. Attach a thin, 5-inch strip of tissue (one that responds to light movement of air currents) to the end of a pencil. Start the engine and run it at normal operating speed. Bring the paper strip close to the various areas of the heat radiating fins and note its behavior. Then bring it close to the flywheel blower and note the movement of the strip. A flow of air currents should be clearly demonstrated. Be careful not to get too close to moving parts of the engine or allow bits of tissue to interfere with the system.

Disassembly Procedure

1. Remove the air shroud, grass screen, recoil starter, etc.
2. Examine the air shroud for cleanliness and any possible damage.
3. Examine the flywheel air vanes for cleanliness and any possible damage.
4. Examine the heat radiating fins for cleanliness and any possible damage.

Part	Disassembly (nuts, bolts, etc.)	Operation performed	Tool used

Reassembly Procedure

Reverse the disassembly procedure.

REVIEW QUESTIONS

Study the air-cooling system and answer the following questions.

1. Trace the airflow through the engine. Make a small sketch.

2. Explain the function of each of the three main parts of the air-cooling system.

3. Explain how to clean heat radiating fins.

4. How often should the system be cleaned?

5. How can the system be damaged?

UNIT 7 IGNITION SYSTEMS

OBJECTIVES

After completing this unit the student will be able to:

- discuss the basic parts of the magneto—how the parts are constructed and how they are mounted on the engine.
- explain the complete magneto cycle.
- list the advantages of the magneto ignition.

Small gasoline engines normally use a magneto for supplying the ignition spark. A *magneto,* figure 7-1, is a self-contained unit that produces the spark for ignition; no outside source of electricity is necessary. It is a simple and very reliable ignition system. Since most small gas engines do not have electric starters, lighting systems, radios, and other electrical accessories, a storage battery is not necessary. The magneto is, therefore, ideally suited for the small gasoline engine.

The following are the basic parts of the magneto ignition system: permanent magnets, high-tension coil (primary and secondary), laminated iron core, breaker points, breaker cam, condenser, spark plug cable, and spark plug. Before discussing how these parts work together, a review is given of some essentials of electricity and magnetism, explaining the construction and function of each individual part.

ELECTRON THEORY

All matter is composed of tiny particles called atoms. The atom is composed of electrons, protons, and neutrons, figure 7-2. The number and arrangement of these particles determines the type of atom: hydrogen, oxygen, carbon, iron, lead, copper, or any other element. Weight, color, density, and all

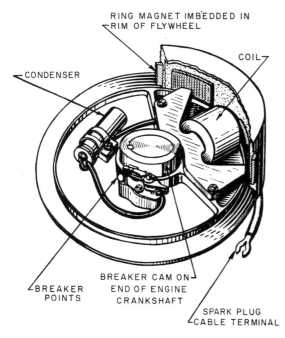

Fig. 7-1 Flywheel magneto

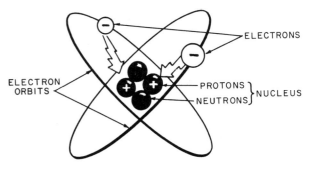

Fig. 7-2 Atomic structure: electron, proton, and neutron

other characteristics of an element are determined by the structure of the atom. Electrons from an atom of copper are the same as electrons from any other element.

The electron is a very light particle that spins around the center of the atom. Electrons move in an orbit. The number of electrons orbiting around the center or nucleus of the atom varies from element to element. The electron has a negative (–) electrical charge.

The proton is a very large and heavy particle in relation to the electron. One or more protons form the center or nucleus of the atom. The proton has a positive (+) electrical charge.

The neutron consists of an electron and proton bound tightly together. Neutrons are located near the center of the atom. The neutron is electrically neutral; it has no electrical charge.

Atoms are normally electrically neutral; that is, the number of electrons and protons are the same, cancelling out each other's electrical force. Atoms stay together because *unlike electrical charges attract each other.* The electrical force of the protons holds the electrons in their orbits. *Like electrical charges repel each other,* so negatively charged electrons do not collide with each other.

In most materials it is very difficult, if not impossible, for electrons to leave their orbit around the atom. Materials of this type are called nonconductors of electricity or *insulators.* Some typical insulating materials are glass, mica, rubber, paper, etc. Electricity cannot flow through these materials.

In order to have electric current, electrons must move from atom to atom. Insulators do not allow this electron movement.

However, in many substances, an electron can jump out of its orbit and begin to orbit in an adjoining or nearby atom. Substances which permit this movement of electrons are called *conductors* of electricity. Some typical examples are copper, aluminum, and silver.

Electron flow in a conductor, figure 7-3, takes place when there is a difference in electrical potential and there is a complete circuit or path for electron flow. In other words, when the source of electricity is short of electrons, it is positively charged. Since unlike charges attract each other, electrons (being negatively charged) move toward the positive source.

A source of electricity can be produced or seen in three basic forms: mechanical, chemical and static. Electricity is produced mechanically in the electrical generator which is commonly connected to water power or steam turbines. The electricity used in homes and factories is produced mechanically. In the magneto, mechanical energy is used to rotate the permanent magnet. Electricity produced by chemical action is seen in the storage battery and dry cell. Static electricity can be seen in nature when lightning strikes. The lightning occurs when the air insulation breaks down and electrons are in a positive area. The lightning may be between clouds, from cloud to earth, or from earth to cloud.

UNITS OF ELECTRICAL MEASUREMENT

There are three basic units of electrical measurement: ampere (rate of electron flow),

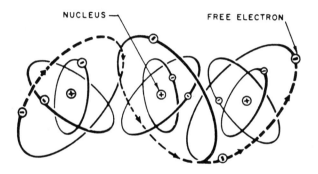

Fig. 7-3 Current: flow of electrons within a conductor

volt (force or pressure causing electron flow), and ohm (resistance to electron flow).

Ampere

The *ampere* is the measurement of electrical current—the number of electrons flowing past a given point in a given length of time. If a person stands at a point on a wire and could count the electrons passing by in one second, 6,250,000,000,000,000,000 electrons would be counted, which equals one ampere of current. To help visualize amperage, think of water flowing in a pipe. A small pipe might deliver two gallons of water a minute. A larger pipe might deliver five gallons of water a minute. Electric wires are generally the same; larger wires can handle more amperage or electron flow than smaller wires.

Volt

The *volt* is the measurement of electrical pressure or the difference in electrical potential that causes electron flow in an electrical circuit. The energy source is short of electrons and the electrons in the circuit want to go to the source. The pressure to satisfy the source is called *voltage*. Voltage might be compared to the pressure that water in a high tank places on the pipe located at the street level. The higher the water pressure, the faster the water flow from a pipe below. Likewise, a higher voltage tends to cause greater flow of electrons.

Ohm

The *ohm* is the unit of electrical resistance. Every substance puts up some resistance to the movement of electrons. Insulators such as porcelain, oils, mica, and glass, for example, put up a great resistance to electron flow. Conductors such as copper, aluminum, and silver put up very little resistance to electron flow. Even though conductors

readily permit the flow of electric current they do tend to put up some resistance. In the water pipe example, this resistance can be seen as the surface drag by the sides of the pipe, or scale and rust in the pipe. Using a larger pipe, or, electrically, a larger wire, is one way of reducing resistance.

OHM'S LAW

In every example of electricity flowing in an electrical circuit, amperes, volts, and ohms each play their part; they are related to each other. This relationship is stated in *Ohm's Law*.

$$\text{Amperes (rate)} = \frac{\text{volts (potential)}}{\text{ohms (resistance)}}$$

The formula is usually abbreviated to:

$$I = \frac{E}{R}$$

For example, if the voltage is 6, and the resistance 12 ohms, the current is calculated as follows.

$$I = \frac{E}{R} \qquad I = \frac{6}{12} \qquad I = .5 \text{ amperes}$$

Of course, the formula can be written to find the resistance, or the voltage.

$$R = \frac{E}{I} \qquad E = I R$$

MAGNETISM

Most people have experimented with a magnet at one time, watching it pick up steel objects and attracting or repelling another magnet. These effects are not entirely explained, but scientists generally agree on the molecular theory of magnetism. Molecules are the smallest divisions of substance that are still recognizable as that substance. Several different atoms may make up one molecule. For example, a molecule of iron oxide contains atoms of iron and oxygen. In many substances the atoms in the molecules are more positive at one spot and more negative at another spot. This is termed a *north pole* and a *south pole.* Usually the poles of adjoining molecules are arranged in a random pattern and there is no magnetic force since their effects cancel one another, as shown in figure 7-4A. However, in some substances, such as iron, nickle, and cobalt, the molecules are able to align themselves so that all north poles point in one direction and all south poles point in the opposite direction, as shown in figure 7-4B. The small magnetic forces of many tiny molecules combine to make a noticeable magnetic force. In magnets, like poles repel each other and unlike poles attract each other, figure 7-5, just as like and unlike electrical charges react. Electricity and magnetism are very closely tied together.

Some substances keep their molecular alignment permanently and are, therefore, classed as *permanent magnets.* Hard steel has this ability. Other materials such as a piece of soft iron (a nail), can attain the molecular alignment of a magnet only when it is

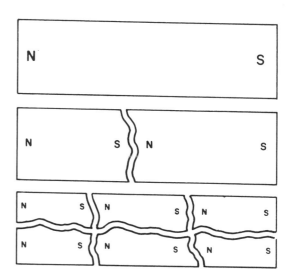

Fig. 7-5 A magnetic field surrounds every magnet. Like poles repel each other; unlike poles attract each other.

in a magnetic field. As soon as soft iron is removed from the magnetic field, its molecules disarrange themselves and the magnetism is lost. Therefore, such substances are termed *temporary magnets.*

More than 100 years ago Michael Faraday discovered that magnetism could produce electricity. Magnetos used on gasoline engines apply this discovery: magnetism producing electricity. Faraday found that if a magnet is moved past a wire, electrical current starts through the wire. If the magnet is stopped near the wire, the current stops. Electricity flows only when the magnetic field or magnetic lines of force are being cut by the wire, figure 7-6.

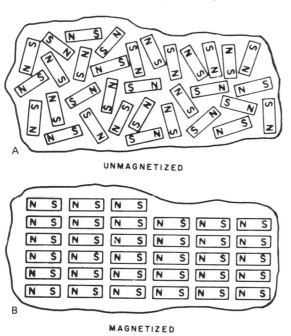

Fig. 7-4 In an unmagnetized bar the molecules are in a random pattern. In a magnetized bar the molecules align their atomic poles.

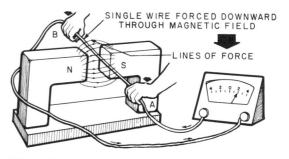

Fig. 7-6 Current flows in the wire as it moves down through the magnetic field.

Another principle that the magneto uses is that when electrons flow through a coil of wire, a magnetic field is set up around the coil, figure 7-7. The coil itself becomes a magnet. Therefore, when electrons flow through the coils in a magneto, a magnetic field is set up.

The principle of the transformer and induced voltage is also used in the magneto. In a transformer, figure 7-8, there is a primary coil and a secondary coil wound on top of the primary; the two are insulated from each other. These coils are wound on a soft iron core. When alternating current passes through the primary coil there is an alternating magnetic field set up in the iron core. The magnetic lines of force cut the secondary coil and induce an alternating voltage within the coil. The voltage produced depends on the ratio of windings in the primary coil and secondary coil. If there are more windings in the secondary than in the primary, the secondary voltage is higher; a *step-up transformer*. If there are more windings in the primary than in the secondary, the secondary voltage is smaller; a *step-down transformer*. Although the magneto does not operate on alternating current, it does use the principle of the step-up transformer.

In a small gasoline engine magneto, there is a magnetic field which induces current in the primary coil, thus setting up a magnetic field around both the primary and secondary coils. At the point of maximum current, the circuit is broken in the primary. Electrons can no longer flow; therefore, the magnetic field collapses. This rapidly collapsing magnetic field induces a very high voltage, igniting the fuel mixture.

Before studying in detail how the magneto operates, it is useful to understand the function of each part. The seven basic parts which work together to produce the ignition spark are the permanent magnets, high-tension coil, laminated iron core, breaker points, condenser, spark plug lead, and spark plug. An example of a magneto ignition system is shown in figure 7-9.

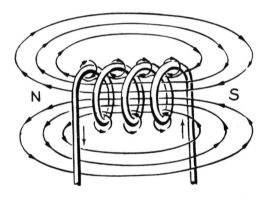

Fig. 7-7 When electricity flows through a coil of wire, a magnetic field is set up around the coil.

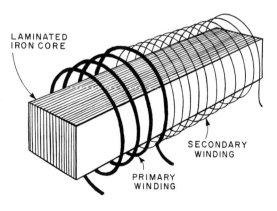

Fig. 7-8 A simple transformer

PERMANENT MAGNETS

Permanent magnets are usually made of an alloy called alnico, a combination of aluminum, nickle, and cobalt. This magnet is quite strong and keeps its magnetism for a very long time. On a flywheel magneto the magnet is cast into the flywheel and cannot be removed. The other magneto parts are often mounted on a fixed plate underneath the flywheel. The magnet, therefore, revolves around the other parts of the magneto. Sometimes the other magneto parts are mounted near the outside rim of the flywheel and the permanent magnets pass by each revolution.

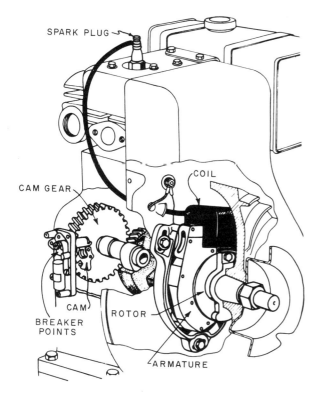

Fig. 7-9 The basic parts of a magneto ignition system

Rotor-type permanent magnets, figure 7-10, are also in use. With this type of construction the permanent magnet rotor may be mounted on the end of the crankshaft or the rotor may be geared to the crankshaft. The rotor lies within the other magneto parts. Care should be taken not to drop or pound the magnet as this causes it to lose some of its magnetism.

HIGH-TENSION COIL (PRIMARY AND SECONDARY WINDINGS)

Refer to figure 7-11. The primary winding of wire consists of about 200 turns of

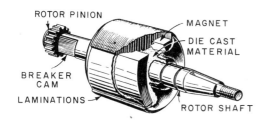

Fig. 7-10 Rotor-type permanent magneto

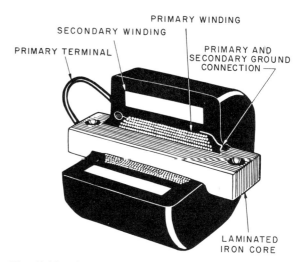

Fig. 7-11 Construction of a high-tension magneto ignition system

a heavy wire (about 18 gauge) wrapped around a laminated iron core. The primary coil is in the electrical circuit containing the breaker points and condenser, figure 7-12. When a magnet is brought near this coil and iron core, magnetic lines of force cut the coil and electrical current is produced. Usually this current flows from the coil, through the

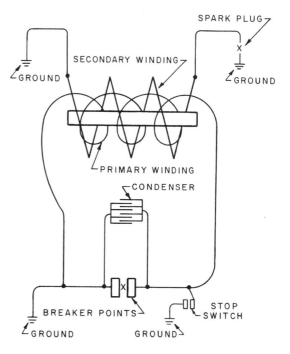

Fig. 7-12 Schematic wiring diagram of a magneto

closed breaker points and is grounded; a complete circuit.

The secondary winding of wire consists of about 20,000 turns of a very fine wire wrapped around the primary coil. The secondary coil is in the electrical circuit containing the spark plug. When current flows in the primary, a magnetic field gradually expands around both coils, but voltages produced in the secondary are quite small. However, when the breaker points open, the circuit is broken, electricity stops flowing and the magnetic field suddenly collapses. The suddenly collapsing magnetic field induces very high voltage in the secondary coil, enough to jump the spark gap in the spark plug.

LAMINATED IRON CORE

The laminated iron core, figure 7-13, is made of many strips of soft iron fastened tightly together. The soft iron core helps to strengthen the magnetic field around the primary and secondary coils, but the core cannot retain the magnetism and become permanently magnetized. The purpose of using many strips instead of a solid core is to reduce eddy currents which create heat in the core. The general shape of the laminated core may vary from magneto to magneto but its function remains the same.

BREAKER POINTS

The breaker points, figure 7-14, are made of tungsten and mounted on brackets. The two points are usually closed or touching each other, providing a path for electron flow. However, just before the spark is desired, the breaker points are opened, breaking the electrical circuit. Breaker points open and close anywhere from 800 times per minute to 4500 or more times per minute, depending on the engine speed.

When the breaker points are open they are usually separated by 0.020 of an inch; the exact opening varies from magneto to magneto. This separation is critical to the function of the magneto; therefore, the breaker point gap must be adjusted correctly.

BREAKER CAM

The breaker cam actuates the breaker points. One bracket of the breaker assembly rides on this cam. As the cam rotates, it opens and closes the breaker points.

On most two-cycle engines, this cam is mounted on the crankshaft and opens the normally closed breaker points once each revolution. On most four-cycle engines, it is mounted to operate from the camshaft which is turning at one-half crankshaft speed. By mounting the breaker points here, they open once every two revolutions of the

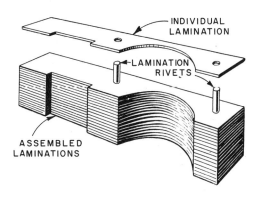

Fig. 7-13 Laminated iron core

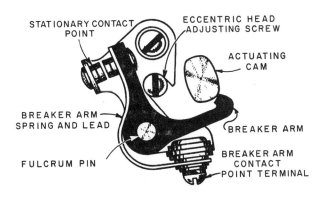

Fig. 7-14 Breaker point assembly

crankshaft. This provides a spark for the power stroke but none for the exhaust stroke. The shape of the cam, figure 7-15, depends on the number of cylinders and the design of the magneto.

CONDENSER

The condenser, figures 7-16 and 7-17, acts as an electrical storage tank in the primary circuit. When the breaker points open quickly, the electrons tend to keep flowing and, if no condenser were present, a spark might actually jump across the breaker points. If this happened, the breaker points would soon burn up, and there would be a weakening effect on the voltage produced by the magneto secondary coil. The condenser provides an electrical storage tank for this last surge of electron flow. The action of the condenser provides for a very abrupt interruption of the primary circuit. The conden-

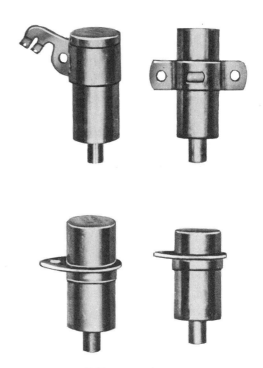

Fig. 7-17 Typical condensers

ser's discharge helps to induce high voltages in the secondary coil as the magnetic fields reverse. The condenser can be easily located because it usually looks like a miniature tin can.

SPARK PLUG CABLE

The spark plug cable connects the secondary coil and the spark plug, providing a path for the high-tension voltage. This part is often referred to as the high-tension lead.

SPARK PLUG

The spark plug is a vital part of the ignition system. In this part the resulting work of the magneto parts is seen. Basically, the spark plug consists of a shell, ceramic insulator, center electrode, and ground electrode. The two electrodes are separated by a gap of about 0.035 of an inch. The path of electricity is down the center electrode, across the air gap, to the ground electrode.

| ONE LOBE CAM | TWO LOBE CAM | THREE LOBE CAM | FOUR LOBE CAM | SIX LOBE CAM |

Fig. 7-15 Common breaker cam shapes

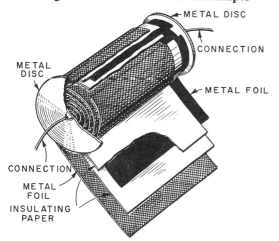

Fig. 7-16 Cutaway showing the construction of a condenser

Of course, the voltage must jump the air gap. When a high-tension voltage of about 20,000 volts is reached, a spark will jump between the electrodes. This spark ignites the fuel mixture within the combustion chamber.

THE COMPLETE MAGNETO CYCLE

Refer to figure 7-18 for illustrations of the steps in the magneto cycle. When the permanent magnet is far away from the high-tension coil, it has no effect on the coil. But as the permanent magnet comes closer and closer, the primary coil feels the increasing magnetic field; the coil is being cut by magnetic lines of force, and, therefore, electrons flow in the primary coil. This current passes through the breaker points and into the ground. When the permanent magnet is nearly opposite the high-tension coil the magnetic field around both coils is reaching its peak. Also, the piston is reaching the top of its stroke, compressing the fuel mixture.

At this time, the position of the permanent magnet causes the polarity of the laminated iron core and the coils to reverse direction. This change of direction is momentarily choked or held back by the coil. Then the breaker points open, interrupting the current flow, and allowing the reversal and rapid collapse of the intense magnetic field that had been built up around the primary and secondary coils. Magnetic lines of force are cutting coils very rapidly, and high voltages are induced in the coils. In the secondary coil the voltage may reach 18,000 to 20,000 volts, enough to jump across the spark gap in the spark plug. This spark ignites the fuel mixture and the piston is forced down the cylinder.

SPARK ADVANCE

When fuel burns in the combustion chamber, it does not explode and exert all of its power immediately. A very short period of time is required for the fuel to ignite and reach its full power. True, this time is very short; but in an engine operating at high speeds, this small time lag is important.

For example, if the mixture was not ignited until the piston reached dead center, the piston would already be starting back down the cylinder before the full force of the burning fuel is reached. This results in loss of power.

Therefore, it is necessary to ignite the fuel slightly before the piston reaches the top of its stroke to realize the full force of combustion. Causing the spark to occur earlier in the engine cycle is called *spark advance.*

The spark is advanced more and more as the engine speed increases, because there is less time for combustion to take place. For example, in a four-cycle engine operating at 2,000 RPM, each power stroke takes about 1/64 of a second. If the speed is increased to 4,000 RPM, each power stroke takes only 1/128 of a second. If the speed is doubled, there is only one-half as much time for combustion to take place. Also, high speeds give the engine higher compression and a more explosive mixture. At high speeds the spark jumps the spark gap before the piston reaches the top of its stroke.

At slow speeds the spark is delayed and occurs later in the cycle, slightly before the piston reaches top dead center. At slow speeds there is more time for combustion to take place. Also, compression is lower and not as much explosive mixture is drawn into the combustion chamber.

Regulating spark advance is done by controlling the time that the breaker points open. Advancing the spark can be done automatically or manually; both methods are commonly used on small gasoline engines.

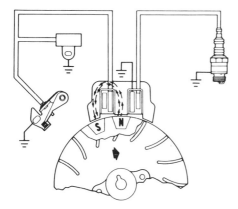

(A) Magnetic field as magnets approach the coil and laminated core

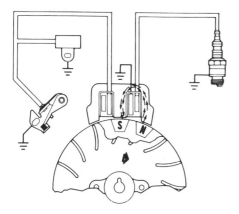

(B) Polarity of magnetic fields reverses rapidly

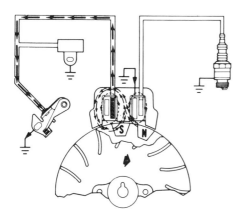

(C) Current flows in the primary circuit as magnetic fields reverse

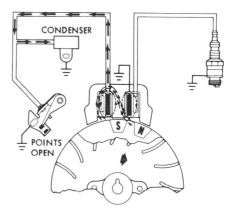

(E) Collapsing magnetic fields and condenser surge induces high voltage to jump the gap at the spark plug

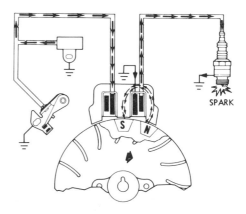

(D) Current stops as breaker points open

Fig. 7-18 The magneto cycle

MANUAL SPARK ADVANCE

Manual spark advance is usually accomplished by loosening the breaker point assembly and rotating it slightly to an advanced position. It is then locked in its new position.

Many outboard motors have a type of manual spark advance, although it is generally referred to by manufacturers as a spark-gas synchronization system. The magneto plate, on which the breaker points are mounted, is located underneath the flywheel. This plate can be rotated through about 25 degrees. The breaker cam is securely mounted on the crankshaft; its relative position cannot be changed. Spark advance is obtained by moving the magneto plate so the breaker points are opened earlier in the cycle. The throttle and magneto plates are linked together so that opening the throttle also advances the spark. At full throttle the spark is fully advanced. At idling speeds the throttle is closed and the magneto plate is positioned for minimum spark advance.

AUTOMATIC SPARK ADVANCE

Automatic spark advance, shown in figures 7-19 and 7-20, is usually accomplished by a centrifugal mechanism which is capable of changing the relative position of the breaker cam and the break points. Here the breaker cam can be rotated through a small distance on its shaft, about 25 degrees. A spring holds the cam in the retarded position at idling and slow speeds but as the engine speed increases, centrifugal force throws the mechanism's hinged weights outward.

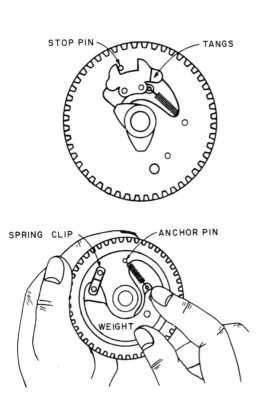

Fig. 7-19 Spark advance mechanism with breaker cam and weight mounted on the camshaft gear

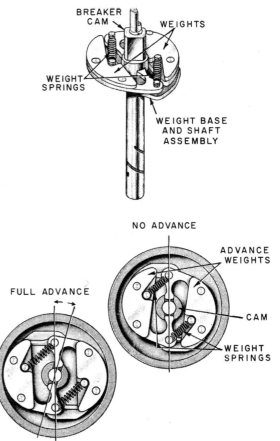

Fig. 7-20 Automatic spark advance mechanism changes the breaker cam's position in relation to the points

This outward motion overcomes the spring's tension and this motion is used to rotate the cam to a more advanced position. At full speed the cam has been rotated its full limit for maximum spark advance.

Some engines, especially larger types, also have a vacuum advance mechanism working with a centrifugal mechanism to provide more accurate spark advance, especially at slow speeds. Automatic spark advance can also be used to move the breaker point assembly to secure the proper advance.

IMPULSE COUPLING

When an engine is started by hand, it is turned over slowly. The voltages produced by the magneto depend upon the speed at which magnetic lines of force cut the primary coil. Therefore, the voltages produced for starting can be quite low, resulting in a weak spark.

Some engines are equipped with an impulse coupling device to supply higher voltages and a hotter spark for slow starting speeds. Using a tight spring and retractable pawls, figure 7-21, the rotation of the magneto's magnetic rotor can be stopped for all

but the last few degrees of its revolution. During the last few degrees of the revolution, the pawl retracts, allowing the spring to snap the magnetic rotor past the firing position at a very high speed. A high voltage can be produced because magnetic lines of force are cutting the primary coil rapidly.

MAGNETO IGNITION FOR MULTICYLINDER ENGINES

Magnetos are often used for multicylinder engines. However, having more than one cylinder to supply with a spark does present a problem. Two solutions to this problem are in common use today.

Two-, three- and four-cylinder engines can have magneto ignition by simply installing a separate magneto for each cylinder. This method is commonly used with outboard engines using a flywheel magneto. Instead of mounting just one magneto on the armature plate under the flywheel, a magneto is installed for each cylinder, figure 7-22. The breaker points of each magneto ride on the common cam. In two-cylinder engines the magnetos are 180 degrees apart, three-cylinder

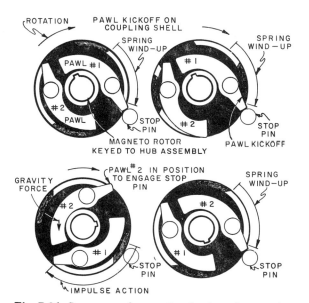

Fig. 7-21 Sequence of operation for impulse coupling for 180 degrees spark magneto

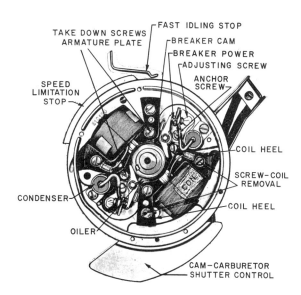

Fig. 7-22 Two complete magnetos are installed on this armature plate.

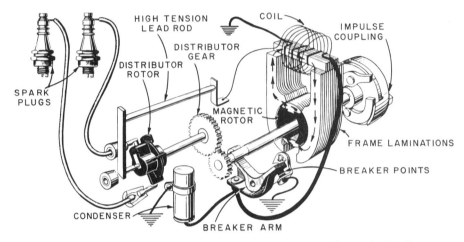

Fig. 7-23 Diagram of rotating-magnet magneto with jump-spark distributor

engines 120 degrees apart, and four-cylinder engines 90 degrees apart. Each magneto functions separately to supply its cylinder with a spark.

Another solution is to use a distributor and two-pole or four-pole magnetic rotor, figure 7-23. Only one complete magneto is used even though the engine may have two or four cylinders. In the case of a two-pole rotor used on a two-cylinder engine, the magneto can produce two sparks every revolution of the magnetic rotor, one every 180 degrees rotation. A two-lobed cam is used to open the breaker points twice each rotor revolution. The distributor rotor channels the high-tension voltage to the correct spark plug.

ENGINE TIMING

The spark must jump the spark gap at exactly the right time, just before the piston reaches top dead center. In many engines, the breaker cam is driven by the camshaft, therefore, the gear on the crankshaft and the camshaft gear must be assembled correctly. These two gears are marked in some manner, figure 7-24, usually with punch marks, so the repairer can easily make the correct assembly. Incorrect alignment of the gears can cause poor operation or no operation at all.

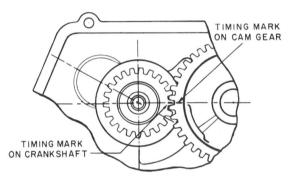

Fig. 7-24 Timing marks on crankshaft gear and cam gear must be aligned correctly.

SOLID STATE IGNITION

Solid state ignition refers to the fact that solid state electronic parts (namely transistors), replace the breaker points. The breaker points in a conventional ignition system can be a source of trouble. Sometimes the systems are called *capacitor discharge systems*, *breakerless ignition* (figure 7-25), or *transistorized ignition*. These systems can be used with just a conventional flywheel magnet as the source of the magnetic force, or they may be used with a generator or an alternator battery system.

The main components of the solid state ignition system, figures 7-26 and 7-27, are a generator or alternator coil, trigger module, ignition coil assembly, and special flywheel

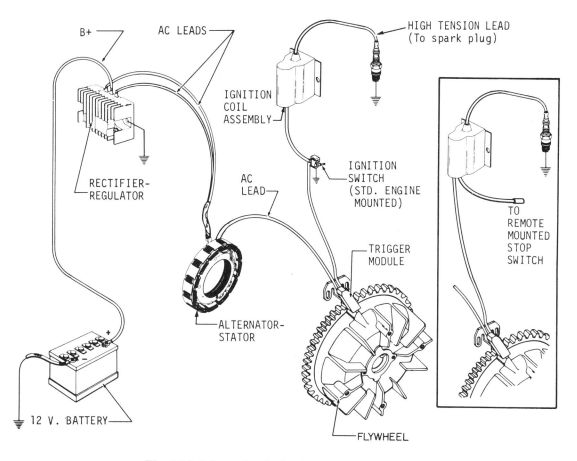

Fig. 7-25 Schematic of a breakerless-alternator system

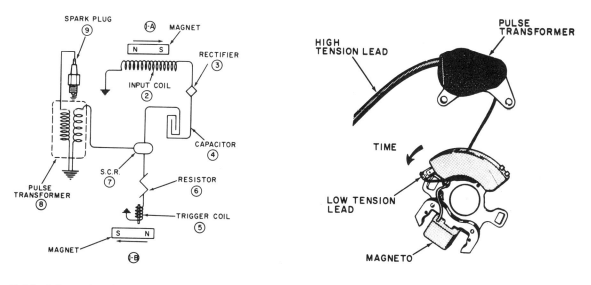

Fig. 7-26 Schematic of a solid state ignition system

Fig. 7-27 Solid state ignition system

with trigger projection. The system has a conventional spark plug and lead.

The trigger module contains transistor diodes which rectify the alternating current, changing it into direct current. Transistors control the flow of electric current by acting somewhat like a valve. Their resistance to flow can be changed by a small current to the transistors. In addition to the diode rectifiers, the trigger module contains a resistor, a sensing coil and magnet, and a silicon-controlled rectifier (SCR). The SCR acts as a switch.

The ignition coil contains primary and secondary windings similar to those of a conventional magneto coil plus a condenser or capacitor. Capacitors and condensers are basically the same, electrically. The operation of the solid· state ignition system follows this cycle:

1. The rotating magnet on the flywheel sets up an alternating current in the alternator or generator coil. This alternating current is rectified into direct current and stored in the capacitor. The diode rectifiers permit flow in one direction only so the capacitor cannot discharge back through the diodes.

2. The magnet group passes the trigger coil, setting up a small current which triggers or gates the SCR. This gating makes the SCR conductive so that the stored voltage of the capacitor surges from the capacitor through the SCR. It then is applied across the primary winding of the ignition coil to the negative side of the capacitor. This surge of energy sets up a magnetic field in the primary winding of the ignition coil. The magnetic field is also induced around the secondary coil, and voltages sufficient to jump the spark gap at the spark plug are reached.

Solid state ignition offers several advantages, such as automatic retarding of the spark at starting speeds, longer spark plug life, faster voltage rise, and a high-energy spark which makes the condition of the plug and its gap less critical.

SPARK PLUG

The spark plug, figure 7-28, is the part of the ignition system that ignites the fuel mixture. It operates under severe and varying temperature conditions and is a critical part in engine operation.

All spark plugs are basically the same but they do differ in these respects: thread size, reach, heat range, and spark gap. There are hundreds of different types of spark plugs, each designed for the special requirements of a certain engine.

The *shell* of the spark plug is threaded so that it can easily be installed or removed from the cylinder head. Various spark plugs have different thread sizes. Some of the more common standard thread sizes are 7/8 inch, 10 mm, 14 mm, and 18 mm.

The *reach* of the spark plug is the distance between the gasket seat and the bottom

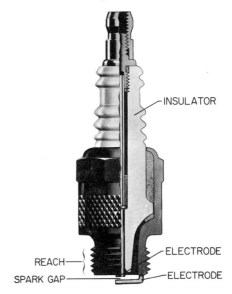

Fig. 7-28 Cutaway of a spark plug

of the spark plug shell, or about the length of the cut threads. The reach ranges from about 1/4 inch to 3/4 inch. Each engine must be equipped with a spark plug of the correct reach since reach determines how far the electrodes protrude into the combustion chamber. If the reach is too small, the electrodes have difficulty igniting the fuel when the spark jumps. If the reach is too great, the top of the piston may strike the electrodes on its upward stroke.

The *heat range* of a spark plug is the range of temperature within which the spark plug is designed to operate. If a spark plug operates at too low a temperature, it is quickly fouled with oil and carbon. If the spark plug operates at too high a temperature, the electrodes quickly burn up. The heat range for spark plugs depends on how fast the heat can be carried away from the electrodes. The path of heat transfer is from the electrodes to the ceramic insulator through the shell and into the cylinder head. The length of the ceramic insulator exposed to the combustion chamber determines the heat range of a spark plug. The longer this insulator, the longer the heat path, and the hotter the spark plug's operating temperature. Likewise, the shorter this insulator, the shorter the heat path, and the colder the spark plug's operating temperature. See figure 7-29.

The *spark gap,* or distance between the electrodes, must be correctly set. It is within this small space that the spark jumps and combustion begins. The gap must be large enough for sufficient fuel mixture to get between the electrodes but not so large as to prevent the spark from jumping across. The spark gap for various plugs ranges from about 0.020 inch to 0.040 inch.

If the spark plug is removed for cleaning, the gap should be checked and reset according to the manufacturer's specifications. A wire gauge should be used, figure 7-30.

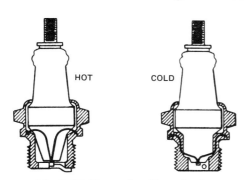

Fig. 7-29 Hot and cold spark plugs

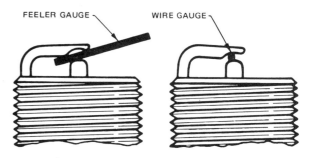

Fig. 7-30 A plain flat feeler gauge cannot accurately measure the true width of a spark gap.

REVIEW QUESTIONS

1. What three particles make up the atom?

2. Explain how electrons can move in a conductor.

3. What are the three ways that electricity can be produced or seen?

4. Explain the following electrical measurements: amperes, volts, and ohms.

5. State Ohm's Law.

6. Explain how permanent and temporary magnets differ.

7. What happens when a wire is cut by magnetic lines of force?

8. What happens to a coil of wire when electric current flows through it?

9. Explain how a transformer works.

10. In a magneto, what is the source of the magnetic field?

11. What are the two parts of the high-tension coil?

12. What purpose does the laminated iron core serve?

13. What is the function of the breaker points? How are they opened?

14. What is the function of the condenser? The spark plug?

15. What are the two electrical circuits in the magneto?

16. Why must the spark be advanced at high speeds?

17. What is the advantage of impulse coupling?

18. Can magneto ignition be used for multicylinder engines?

19. In what four ways do spark plugs differ?

CLASS DISCUSSION TOPICS

- Discuss the structure of an atom.
- Discuss how conductors and insulators differ.
- Discuss the molecular theory of magnetism.
- Discuss the advantages of magneto ignition.
- Discuss the complete magneto cycle.

CLASS DEMONSTRATION TOPICS

- Demonstrate a magnetic field around a magnet using a permanent magnet, iron filings, and paper.
- Demonstrate current flow caused by varying magnetic field using a galvanometer, coil of wire, and permanent magnet.
- Demonstrate a magnetic field around a coil of wire using a direct-current source, coil of wire, iron filings, and paper.
- Disassemble a magneto showing the basic parts.
- Demonstrate how the breaker points are opened and show their separation at maximum opening (feeler gauge).

LABORATORY EXPERIENCE 7-1
BASIC FLYWHEEL MAGNETO PARTS

The engine is disassembled to expose flywheel magneto parts. The arrangement of parts is studied and the engine is turned over so the breaker points can be observed opening and closing. Basic parts of the ignition system are removed for closer inspection. Students are to record their work after each step of the disassembly procedure is completed.

Disassembly Procedure

Instructors may want to supplement or revise specific steps of this procedure since there are many makes of engines. The following procedure is a general guide.

1. Remove the spark plug lead from the spark plug.
2. Remove any air shrouding, grass screens, or recoil starters, from the flywheel area.
3. Remove the flywheel and flywheel nut. Use a flywheel puller if one is available. If a flywheel puller is not at hand, deliver a sharp blow to the end of the crankshaft with a plastic or soft hammer.
4. Remove the breaker point cover (on some engines).
5. Turn the crankshaft over and observe the opening and closing of the breaker points.
6. Remove the high-tension coil and laminated iron core.
7. Remove the condenser.
8. Remove the breaker point assembly.
9. Remove the breaker cam (on some engines).

Part	Disassembly (nuts, bolts, etc.)	Operation performed	Tool used

Reassembly Procedure

Reverse the disassembly procedure. If the engine is to be operated upon reassembly several points are very important.

1. Reinstall the breaker cam in the same way as it came off.

2. Use care with the key and the keyways; do not force.

3. Breaker points must be reset to manufacturer's specifications.

4. The laminated iron core must be positioned for the proper air gap between it and the flywheel magneto.

5. All connections should be checked for tightness.

REVIEW QUESTIONS

Study the flywheel magneto system and then answer the following questions.

1. List the magneto parts in the primary circuit.

2. List the magneto parts in the secondary circuit.

3. Explain how the spark is triggered.

4. On which engine stroke is the spark provided?

5. What opens and closes the breaker points?

6. What happens to the primary electrical circuit when the breaker points open?

7. What happens to the magnetic field when the breaker points open?

8. What is the purpose of using a laminated iron core instead of a solid core?

9. What are the functions of the condenser?

10. Discuss briefly what happens in the magneto cycle with one revolution of the crankshaft.

UNIT 8 ROUTINE CARE AND MAINTENANCE, AND WINTER STORAGE

OBJECTIVES

After completing this unit the student will be able to:

- clean and adjust spark plugs.
- correctly perform winter storage.
- clean an air cleaner.

A gasoline engine represents an investment. To safeguard this investment the engine operator must perform certain routine steps in care and maintenance. Routine care and maintenance help to ensure the longest life possible for engine parts and can save on repair bills.

The best source of information on engine care is the owner's manual or operator's instruction book, that comes with every new engine. The book contains the information that the manufacturer considers necessary for the owner to know. The information included allows the owner to obtain top performance and long engine life. Some typical topics discussed are lubrication procedures, carburetor adjustment, ignition system data, air cleaner cleaning instructions, winter storage, and care of the machine that the engine is used on. The information in the book is important; a person with a new engine should read and thoroughly understand the instruction book before touching the engine. A plastic cover helps to keep the book in good condition.

ROUTINE CARE AND MAINTENANCE

Routine care and maintenance is given to an engine during its normal use. It is provided to keep the engine operating at its peak efficiency and to prevent undue wear of engine parts. There are four basic points to consider: oil supply, cooling system, spark plug, and air cleaner.

OIL SUPPLY

On a four-cycle engine, the oil supply in the crankcase should be checked daily or each time the engine is used. If the oil level has dropped below the add mark, fill the crankcase to the proper level.

Most manufacturers recommend that the oil be changed after a short period of time when breaking in a new engine. Depending on the manufacturer, this first oil change should be accomplished from two to twenty hours.

After the oil has been changed once, the engine can go for longer periods of time between oil changes. Lauson engines specify changing the oil every ten hours of operation; Briggs and Stratton, Whizzer, and Kohler engines every twenty hours; Gravely and Wisconsin engines every fifty hours; Onan engines after one hundred hours. Each manufacturer's instruction book gives the length of time between oil changes. On two-cycle engines, oil changes are not necessary since all lubricating oil is mixed with the gasoline.

COOLING SYSTEM

The most important consideration regarding the cooling system is to keep it clean. On air-cooled engines, clean the radiating fins (figure 8-1), flywheel vanes, and shrouds whenever they begin to accumulate grass, dirt, etc. Do not allow deposits to build up, reducing the engine's cooling capacity. How often the cooling system is cleaned depends on the dust conditions under which the engine operates.

On a water-cooled engine, check the water level in the radiator before operating the engine. If it is low, fill it to the correct level.

Outboard motors need little routine care for the cooling system. However, each time the engine is started the operator should check to see that the engine is pumping water. This can be detected on most engines by a small amount of water coming from the telltale holes. If the engine is not pumping water, it should be stopped immediately and the source of the trouble corrected. Do not put hands by the underwater discharge to determine if the engine is pumping water, for the propeller may be dangerously close. If the engine's cooling system is thermostatically controlled it may take a few moments for the thermostat to open, allowing a flow of water from the engine.

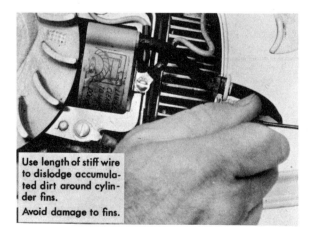

Use length of stiff wire to dislodge accumulated dirt around cylinder fins.

Avoid damage to fins.

Fig. 8-1 Keep heat radiating fins clean.

SPARK PLUGS

Spark plugs need to be cleaned and regapped periodically. Spark plugs are usually cleaned after one hundred hours of operation, but some manufacturers recommend cleaning after as little as fifty hours. Since spark plugs operate under severe conditions, they become fouled easily. Knowing the signs of fouling can often tell about the engine's overall condition. Common problems that develop with spark plugs are carbon fouling, burned electrodes, chipped insulator, and splash fouling. See figure 8-2, page 142.

Normal spark plugs have light tan or gray deposits but show no more than 0.005 inch increase in the original spark gap. These can be cleaned and reinstalled in the engine.

Worn out plugs have tan or gray colored deposits and show electrode wear of 0.008 inch to 0.010 inch greater than the original gap. Throw away such plugs and install new ones.

Oil fouling is indicated by wet, oily deposits. This condition is caused when oil is pumped (by the piston rings) to the combustion chamber. A hotter spark plug can help but the condition may have to be corrected by engine overhaul.

Gas fouling or fuel fouling is indicated by a sooty, black deposit on the insulator tips, electrodes, and shell surfaces. The cause may be a too rich fuel mixture, light loads, or long periods at idle speed.

Carbon fouling is indicated if the plug has dry, fluffy, black deposits. This condition may be caused by a too rich fuel mixture, improper carburetor adjustment, choke partly closed, or clogged air cleaner. Also, slow speeds, light loads, long periods of idling, and the resulting cool operating temperature can cause deposits not to be burned away. A hotter spark plug may correct carbon fouling.

Lead fouling is indicated by a soft, tan, powdery deposit on the plug. These deposits

NORMAL SPARK PLUG

WORN OUT SPARK PLUG

CARBON-FOULED SPARK PLUG

CHIPPED INSULATOR ON SPARK PLUG

OVERHEATED SPARK PLUG

OIL-FOULED SPARK PLUG

SPLASH-FOULED SPARK PLUG

BENT SIDE ELECTRODE

Fig. 8-2 Normal and damaged spark plugs

of lead salts build up during low speeds and light loads. They cause no problem at low speed but at high speeds, when the plug heats up, the fouling often causes the plug to misfire, thus limiting the engine's top performance.

Burned electrodes are indicated by thin, worn away electrodes. This condition is caused by the spark plug overheating. This overheating can be caused by lean fuel mixture, low octane fuel, cooling system failure, or long periods of high-speed heavy load. A colder plug may correct this trouble.

Splash fouling can occur if accumulated cylinder deposits are thrown against the spark plugs. This material can be cleaned from the plug and the plug reinstalled. If spark gap tools or pliers are not properly used, the electrode can be misshaped into a curve.

CLEANING SPARK PLUGS

If inspection indicates that the spark plug needs cleaning, it can be cleaned within a few minutes. First, wire brush the shell and threads, figure 8-3, not the insulator and electrodes. Then wipe the plug with a rag that has been saturated with solvent. This removes any oil film on the plug. Next, file the sparking surfaces of the electrodes with a point file, figure 8-4. Finally, regap the electrodes, figure 8-5, setting them according to specifications. Check the gap with a wire spark gap gauge, figure 8-6. Most service stations have

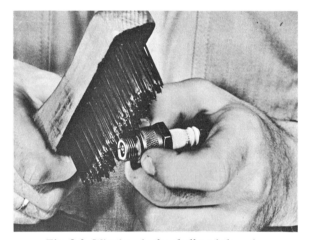

Fig. 8-3 Wire brush the shell and threads.

Fig. 8-5 Regap the plug.

Fig. 8-4 File electrodes.

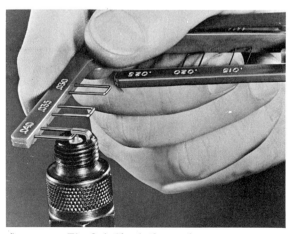

Fig. 8-6 Check the spark gap.

an abrasive blast machine that is good for cleaning around the electrodes and insulator. Do not allow any foreign material to fall into the cylinder while the spark is out. Also, do not forget the spark plug gasket when reinstalling the spark plug. This copper ring provides a perfect seal.

Of course, cleaning cannot repair a broken, cracked, or otherwise severely damaged spark plug. Cleaning is designed to maintain good operation and prolong spark plug life. Many small engine manufacturers recommend the installation of a new spark plug at the beginning of each season. On a single-cylinder engine, the spark plug is relatively more critical than on a six- or eight-cylinder engine.

AIR CLEANER

The air cleaner serves the important function of cleaning the air before it is drawn through the carburetor and into the engine. Small abrasive particles of dust and dirt are trapped in the air cleaner, figure 8-7. The air cleaner should be periodically cleaned; just how often depends mainly on the atmosphere in which the engine is operating. In a dusty atmosphere, (a garden tractor used in a dry garden for example), the air cleaner should be cleaned every few hours. Under normal conditions, the air cleaner should be cleaned every twenty-five hours or sooner.

Fig. 8-7 About this much abrasive material would enter a six-cylinder engine every hour if the air cleaner were not used.

Air cleaners can be classified as either *oil type,* which includes oil-bath air cleaners and oil-wetted polyurethane (foam) filter element; or *dry type,* which includes foil, moss, or hair element, felt or fiber hollow element, and metal cartridge air cleaners.

OIL-TYPE AIR CLEANERS

Oil-bath Air Cleaner. This air cleaner, figure 8-8, carries a small amount of oil in its bowl. The level of this oil should be checked before each use of the engine. If the oil level is low, fill it up to the mark, not above. Use the same type of oil as is used in the crankcase. As this cleaner works, both the oil and filter element become dirty. To clean an oil-bath air cleaner, disassemble the unit and then pour out the dirty oil. Wash the bowl, cover, and filter element in solvent. Dry all parts. Refill the bowl to the correct level and then reassemble the unit, figure 8-9.

Oil-wetted Polyurethane Filter Element Air Cleaner. Shown in figure 8-10, this type of air cleaner is perhaps the most common. Clean it by washing the element in kerosene, liquid detergent and water, or an approved solvent. Dry the element by squeezing it in a cloth or towel. Apply considerable oil to the foam and work it throughout the element. Squeeze out any excess oil and then reassemble the air cleaner.

Fig. 8-8 Oil-bath air cleaner

Fig. 8-9 Clean, fill, reassemble air cleaner

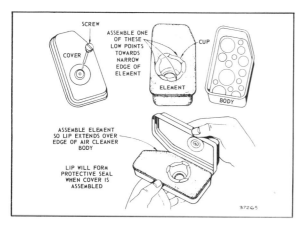

Fig. 8-10 Oil foam air cleaner

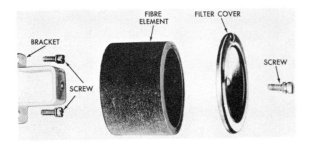

Fig. 8-11 Fiber-element type air cleaner

DRY-TYPE AIR CLEANERS

Foil, Moss, or Hair Element Air Cleaner. This type of element is also washed in solvent. After washing, allow the element to dry, then return it to the air cleaner body.

Felt or Fiber Cylinder Element Air Cleaners. These air cleaners should be cleaned by blowing compressed air through them in the opposite direction from normal flow. Dirt and dust particles will be blown out of the element.

Metal-Cartridge Air Cleaners. These air cleaners, shown in figure 8-12, page 146, are cleaned by tapping or shaking to dislodge dirt accumulations.

Air cleaner elements are replaceable parts. If they wear out or become very clogged, they should be thrown away and replaced with a new filter element. As a simple test to determine if the air cleaner is badly clogged: (1) run the engine with the air cleaner removed; (2) then, with the engine still running, replace the air cleaner; and (3) notice if the engine speed remained constant or dropped down. Any noticeable drop in speed probably indicates that the filter element is clogged.

Periodically inspect the engine for loose parts and do not neglect lubricating the engine's associated linkages, gear reduction units, chains, shafts, wheels, or the machinery the engine powers. These requirements vary greatly from engine to engine. On an outboard engine remember that the lower unit requires special lubrication.

EXHAUST PORTS (2 CYCLE)

Cleaning the exhaust ports of a two-cycle engine, figure 8-13, page 146, can also be considered a part of routine care and maintenance. Over a period of time these ports can become clogged with carbon deposits. These deposits can greatly reduce power and performance. After removing

Fig. 8-12 Metal-cartridge type air cleaner

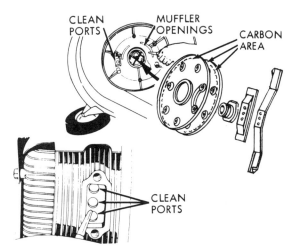

Fig. 8-13 Cleaning the exhaust ports

the muffler to expose the ports, the carbon can be cleaned out with a soft object such as a dowel rod. Be sure to move the piston so that it covers the ports to prevent loose deposits from going into the cylinder.

WINTER STORAGE AND CARE

Putting up an engine at the end of the season should be done carefully. These are several items that should be done:

- Drain the fuel system.
- Inject oil into the cylinder (upper cylinder).
- Drain the crankcase (four cycle).
- Clean the engine.
- Wrap the engine in a canvas or blanket and store in a dry place.

The entire fuel system should be drained; tank, fuel lines, sediment bowl, and carburetor. Gasoline that is stored over a long period of time has a tendency to form gummy deposits or varnish. These deposits could injure and block the fuel system if gasoline stands in the system for a long time. Do not plan to hold gasoline over from season to season. Purchase a new supply when the engine is returned to use.

Many manufacturers recommend that a small amount of oil be poured into the cylinder prior to storage. Remove the spark plug and pour in the oil (about a tablespoon in most cases), then turn the engine over by hand a few times to spread the oil evenly over the cylinder walls. Finally, replace the spark plug.

Some manufacturers suggest that the oil be introduced into the cylinder while the engine is running at a slow speed and just before the engine is stopped. To do this, the air cleaner is removed and a small amount of oil is poured in. The oil goes through the carburetor and then into the combustion chamber. When dark blue exhaust is produced the engine can be stopped.

The crankcase should be drained while the engine is warm so the oil drains more readily. In some cases the crankcase is then refilled but on most engines it need not be refilled until the engine is returned to service.

Any unpainted surfaces that might rust should be lightly coated with oil before storage. Linkages should be oiled. The radiating fins and the engine in general should be cleaned. Do not put too much oil on the engine as it will collect dirt and be difficult to remove.

Finally, wrap the engine in a canvas or dry blanket and store the engine in a dry place. A garage, dry shed, or dry basement is good. Do not allow the engine to be left outside, exposed to the weather.

RETURNING THE ENGINE TO SERVICE

Upon returning the engine to service (see figure 8-14), refill the gas tank with new gasoline. Before filling the gas tank check to see if any water has condensed in the tank. If so, drain the water completely.

Check for condensation in the crankcase and refill the crankcase with new oil of the

(A) Fill the gas tank.

(D) Choke closed

(B) Check the oil in the crankcase.

(E) Spark plug shorting bar off spark plug

(C) Fuel shut-off valve open

(F) Air cleaner clean

Fig. 8-14 Things to check before returning the engine to service

correct type. Also, clean the air cleaner and refill the bowl with oil if it is an oil-bath type.

Check the spark plug; clean and re-gap it. Many manufacturers of small engines recommend that a new plug be installed at the beginning of each season to ensure peak performance throughout the season.

SPECIAL CONSIDERATIONS FOR OUTBOARD MOTORS

Winter Storage

Generally, winter storage for outboard motors is similar to that for other engines. One notable difference is the cooling system. Being water cooled, all water must be removed from the cooling system. If a water-filled cooling system on an idle engine is exposed to freezing weather, cracked water jackets and other major damage can result.

To remove the water from the system, remove the engine from the water, set the speed control on stop (to prevent accidental starting), and turn the engine over several times by hand. This allows the water to drain from the water jackets and passages.

Salt Water Operation

Although most engines are treated with anticorrosives, not all engines are free from corrosion. If an engine normally used in salt water is not to be used for a while, its cooling system should be flushed with fresh water to prevent any possible corrosion.

Use in Freezing Temperatures

If the engine is being used during freezing temperatures, care must be taken not to allow the cooling water to freeze in the engine or lower unit during an idle period. Of course while the engine is operating there is no danger of freezing.

REVIEW QUESTIONS

1. Explain what routine care and maintenance is and why it is important.

2. Where can routine care and maintenance information be found?

3. What are the four basic points to be considered in routine care and maintenance?

4. How often should the oil in the crankcase be checked?

5. How often should the oil in the crankcase be changed (Briggs and Stratton engine for example)?

6. What are the two types of cooling systems?

7. How often should the air-cooling system be cleaned?

8. How is an air-cooling system cleaned?

9. What care should be given a water-cooling system?

10. How often should spark plugs be cleaned?

11. What is the purpose of cleaning a spark plug?

12. List several types of spark plug fouling.

13. What are the steps in cleaning a spark plug?

14. What is the purpose of the air cleaner?

15. How often should air cleaners be cleaned?

16. What are the two basic types of air cleaners?

17. Discuss the cleaning of the several types of air cleaners.

18. List the general steps in winter storage of engines.

19. List the general steps in returning the engine to service.

20. List four safety rules for gasoline engines.

21. What special considerations must be made for outboard motors?

CLASS DISCUSSION TOPICS

- Discuss the importance of routine care and maintenance.
- Discuss the importance of care in winter storage.
- Discuss the types of cooling systems and their maintenance.
- Discuss the types of spark plug fouling.
- Discuss the importance and function of the air cleaner.
- Discuss the various types of air cleaners and how each might be used.
- Discuss the importance of safety with gasoline and gasoline engines.
- Discuss state laws in regard to gasoline storage.

CLASS DEMONSTRATION TOPICS

- Demonstrate cleaning and care of cooling systems.
- Demonstrate cleaning spark plugs.
- Demonstrate cleaning air cleaners.
- Demonstrate putting an engine into winter storage.
- Demonstrate returning an engine to service after winter storage.
- Demonstrate the check-out of an engine prior to starting.
- Demonstrate routine care and maintenance on an outboard engine.
- Demonstrate routine care and maintenance on a four-stroke cycle engine.

LABORATORY EXPERIENCE 8-1
CLEAN AND REGAP SPARK PLUGS

The engine spark plug is removed, examined for damage and fouling, cleaned, and re-gapped. Students are to record their work after each step of the procedure is completed.

If spark gap tools and/or pliers are not properly used the electrode can be misshaped into a curve.

Correct spark plug type and correct spark plug gap setting are found in the engine operator's manual.

Procedure

1. Remove the spark plug and examine the condition of the plug.

2. Check the spark gap before cleaning and resetting.

3. Wire brush the shell and threads.

4. Clean the insulator with a rag soaked in solvent.

5. Clean electrodes and sparking surfaces.

6. Regap the plug to the correct setting.

7. Reinstall the plug. (Do not forget the spark plug gasket.)

Part	Disassembly (nuts, bolts, etc.)	Operation performed	Tool used

REVIEW QUESTIONS

Upon completion of the work, answer the following questions.

1. What is the spark plug type (commercial designation)?

2. What was the spark gap prior to cleaning?

3. Does the manufacturer suggest any equivalent spark plug types? If so, list them.

4. What is the correct spark gap for the plug?

5. Describe the condition of the plug prior to cleaning.

6. Explain the purpose of cleaning spark plugs.

LABORATORY EXPERIENCE 8-2
WINTER STORAGE OF ENGINES

An engine is prepared for winter storage (either two- or four-cycle). Students are to record their work after each step of the procedure is completed.

Procedure

Instructors may want to supplement or revise specific steps of this procedure since there are many makes of engines. The following procedure is a general guide.

1. Remove the spark plug lead.

2. Drain the fuel system: tank, fuel lines, sediment bowl, and carburetor. Clean the air cleaner.

3. Inject oil into the cylinder (upper cylinder). Remove the spark plug and pour in about one ounce of oil. Turn the engine over several times to spread the oil.

4. Drain the crankcase (four cycle).

5. Clean the engine. Pay special attention to the cooling system.

6. Wrap the engine in a canvas or blanket and store in a dry place.

Part	Disassembly (nuts, bolts, etc.)	Operation performed	Tool used

REVIEW QUESTIONS

Answer the following questions.

1. Explain why winter storage is important.

2. Should gasoline be kept over for the next season? If not, why?

3. Is winter storage a job that can be done by an engine owner?

4. Why should the air cleaner be cleaned before storage instead of when the engine is returned to service?

5. List several good places to store an engine.

LABORATORY EXPERIENCE 8-3
INSTRUCTION BOOK USAGE

Students are to read an instruction book (one of their own, or one the instructor gives them), and then complete the information requested below. Every manufacturer has a slightly different idea of just how much information the engine owner should have to operate the engine. Therefore, it may not be possible for each student to complete all the information.

Note: See figure 8-15 for a sample engine identification.

Engine Manufacturer's Name

Engine Model Number

Two- or Four-cycle Engine

Rated Horsepower   at   RPM

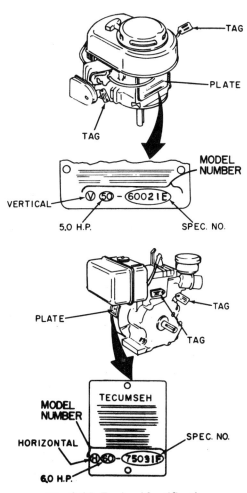

Fig. 8-15 Engine identification

Bore   Stroke   Displacement

Lubrication System:

    Summer Oil Type

    Winter Oil Type

    Oil Checked — How Often

    Oil Changed — How Often

Ignition System:

    Spark Plug Type

    Spark Plug Gap

    Spark Plug Cleaned — How Often

    Breaker Points Checked — How Often

    Breaker Points Set To

Air Cleaner:

    Dry Type or Oil Bath

    Cleaned — How Often

    Air Cleaner Cleaned With

Cooling System:

    Air-Cooled or Water-Cooled

    Cleaned — How Often (Air-Cooled)

    Water Level Checked (Water-Cooled) — How Often

    NOTE: Most questions are answered in terms of hours between cleaning, changing, setting, checking, and so forth.

LABORATORY EXPERIENCE 8-4
CLEANING THE AIR CLEANER

The air cleaner is disassembled, inspected, cleaned, and reassembled. Students are to record their work after the work is completed.

Cleaning Procedure

The engine instruction book is the final word on the correct cleaning of the air cleaner. If an instruction book is not available, the following procedure can be used as a general guide. Be careful not to allow any dirt to fall into the carburetor when the air cleaner is removed.

A. Oil-Bath Air Cleaner

   1. Loosen and carefully remove the air cleaner from the carburetor.

   2. Wash out the filter element in solvent.

   3. Pour the old oil out of the bowl, and wipe the bowl clean.

   4. Refill the bowl to the correct level. (Use the same type oil as that in the crankcase.)

   5. Reassemble the air cleaner.

B. Oil-Wetted Polyurethane (Foam) Filter Element (See figure 8-16.)

   1. Loosen and carefully remove the air cleaner from the carburetor.

   2. Wash out the filter element in solvent.

   3. Squeeze out excess solvent.

   4. Work about one teaspoon of oil into the filter.

   5. Reassemble the air cleaner.

C. Foil, Moss, or Hair Elements; Dry Type

   1. Loosen and carefully remove the air cleaner from the carburetor.

   2. Wash out the filter element in solvent, and allow it to dry.

   3. Reassemble the air cleaner.

D. Felt or Fiber Cylinder Element; Dry Type

   1. Loosen and carefully remove the air cleaner from the carburetor.

   2. Blow out the element with compressed air, and reverse direction of the normal flow.

   3. Reassemble the air cleaner.

E. Metal Cartridge Air Cleaner; Dry Type

   1. Loosen and carefully remove the air cleaner from the carburetor.

   2. Tap or shake the air cleaner to dislodge dirt and dust.

   3. Reassemble the air cleaner.

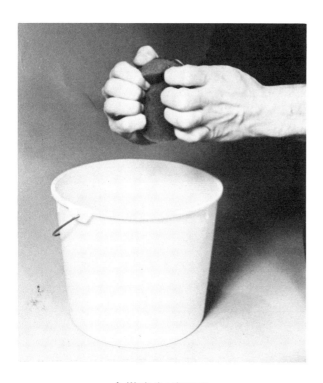

A Wash the element

C Pour oil on the element

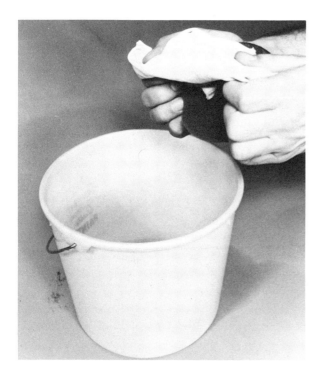

B Dry the element

D Work oil into the element

Fig. 8-16 Cleaning polyurethane foam air cleaner

Part	Disassembly (nuts, bolts, etc.)	Operation performed	Tool used

REVIEW QUESTIONS

Upon completion of the work, answer the following questions.

1. Explain what would happen to the fuel mixture if the air cleaner gradually became clogged.

2. What harmful effect can dirty air have on an engine?

3. Classify the following operations in relation to atmospheric conditions (dusty, average, or clean):

   a. Engine operated in feed mill

   b. Engine on small snow plow

   c. Engine mulching dry leaves

   d. Outboard motor on lake

   e. Engine on garden tractor

UNIT 9 TROUBLESHOOTING, TUNE-UP, AND RECONDITIONING

OBJECTIVES

After completing this unit the student will be able to:

- correctly clean and adjust the breaker points, and test the spark produced at the spark plug.
- test the magneto output, and correctly adjust engine timing.
- correctly replace piston rings, and correctly lap valves.
- check the compression of an engine.
- correctly use the feeler gauge and torque wrench, and perform an engine tune-up.

Troubleshooting is the intelligent, step-by-step process of locating engine trouble. The troubleshooter examines and/or tests the engine to determine the cause of its disorder. To engage in this process a person must thoroughly understand the how and why of engine operation. Troubleshooting is more a mental activity than a physical one.

The troubleshooter first establishes the engine's symptoms. The most common causes for these symptoms are then checked. If the solution is not found, another possible cause for the engine's problem is explored. The possible cause for engine failure must quickly be narrowed down, because there are many afflictions that can creep into an ailing engine.

An experienced mechanic troubleshoots quickly and with little apparent effort. Years of experience and a thorough understanding of engine operation allow the mechanic to work with such competence.

Engine owners can do their own troubleshooting when they understand engine operating principles. Many owner's manuals contain a troubleshooting chart that is of great assistance. With the aid of an engine trouble-shooting chart, the engine owner or operator can locate and correct many engine difficulties without calling in a professional mechanic.

Troubleshooting requires a knowledge of engine operation, as well as some common sense. For example, if the engine does not start, there are two possible reasons: either (1) fuel is not getting into the combustion chamber, or (2) fuel is being provided but it is not being ignited. Therefore, the trouble is either in the fuel system or the ignition system. Lubrication or cooling systems have little influence on the ability of the engine to start, so these systems are not inspected.

The troubleshooter starts with the simplest and most frequent causes of trouble. In inspecting the fuel system, the first step is to check for fuel in the tank. If this is not the cause of the trouble, then check the fuel shutoff valve, continuing logical steps until the trouble is found. It would be a mistake to completely disassemble the carburetor or magneto as the first step in troubleshooting.

Look at the troubleshooting charts (figure 9-1, page 162, and in the Appendix) and discover how many of the common

FAULT	POSSIBLE CAUSE
Poor Compression 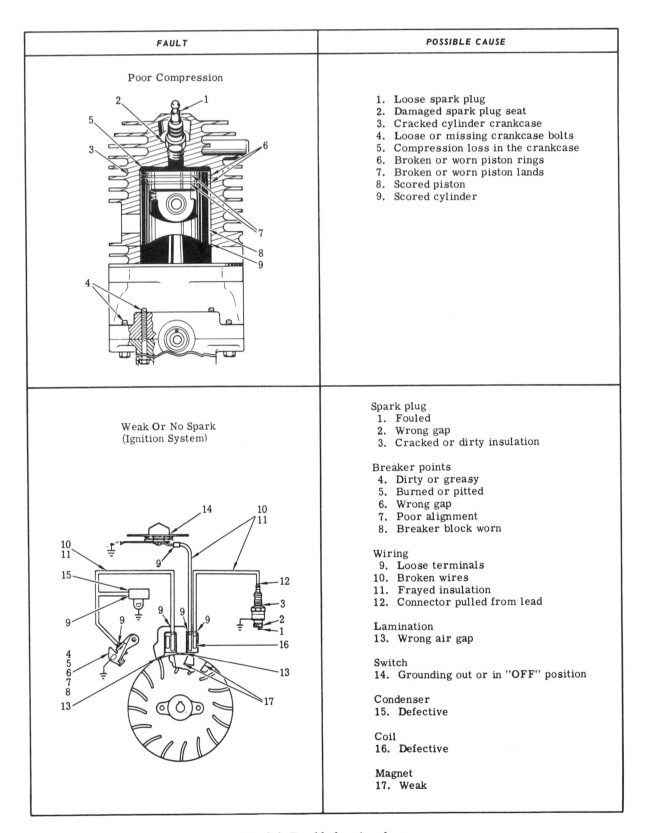	1. Loose spark plug 2. Damaged spark plug seat 3. Cracked cylinder crankcase 4. Loose or missing crankcase bolts 5. Compression loss in the crankcase 6. Broken or worn piston rings 7. Broken or worn piston lands 8. Scored piston 9. Scored cylinder
Weak Or No Spark (Ignition System)	Spark plug 1. Fouled 2. Wrong gap 3. Cracked or dirty insulation Breaker points 4. Dirty or greasy 5. Burned or pitted 6. Wrong gap 7. Poor alignment 8. Breaker block worn Wiring 9. Loose terminals 10. Broken wires 11. Frayed insulation 12. Connector pulled from lead Lamination 13. Wrong air gap Switch 14. Grounding out or in "OFF" position Condenser 15. Defective Coil 16. Defective Magnet 17. Weak

Fig. 9-1 Troubleshooting chart

engine troubles can be corrected by an owner who has the household tools on hand.

ENGINE TUNE-UP

Engine tune-up does not involve major engine repair work. Rather, it is a process of cleaning and adjusting the engine so that it gives top performance. A tune-up can be done by an experienced engine owner or it can be done by a mechanic. A typical tune-up procedure includes these steps.

1. Inspect the air cleaner; then clean and reassemble the air cleaner. Is the cleaner damaged in any way? Is the filter element too clogged for cleaning? If so, replace the cleaner and/or element. Clean the unit according to the manufacturer's instructions.

2. Clean the gas tank, fuel lines, and any fuel filters or screens.

3. Check the compression. This can be done by slowly turning the engine over by hand. As the piston nears top dead center, considerable resistance should be felt. As top dead center is passed, the piston should snap back down the cylinder. A more thorough test can be made with a compression gauge, figure 9-2. The manufacturer's repair manual should be consulted for normal and minimum acceptable compression pressure.

4. Check the spark plug; clean, regap, or replace. Remove the high-tension lead from the plug and hold it about 1/8 inch from the plug base, figure 9-3. Be sure to hold it on the insulation. Turn the engine over. If a good spark jumps to the plug, the magneto is providing sufficient spark. Now replace the high-tension lead and lay the plug on a bare spot on the engine, figure 9-4. Turn the engine over. If a good spark jumps at the plug electrodes, the plug is good.

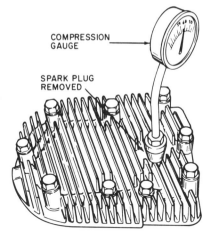

FIRST TEST: WITHOUT OIL IN CYLINDER
SECOND TEST: SQUIRT A FEW DROPS OF OIL ONTO PISTON THROUGH SPARK PLUG HOLE. TURN ENGINE OVER 6 TO 8 REVOLUTIONS TO GET OIL TRANSFERRED TO PISTON RING AREA, THEN MAKE COMPRESSION TEST.

Fig. 9-2 Checking compression with a gauge

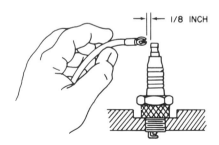

Fig. 9-3 Checking for spark

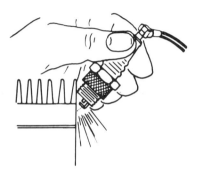

Fig. 9-4 Checking spark plug

This is not an absolute test but is a good indication. If the plug is questionable, replace it.

5. Check the operation of the governor, figure 9-5. Be certain the governor linkages do not bind at any point.

6. Check the magneto. On most engines, the flywheel must be removed for access to the magneto parts. Use a flywheel or gear puller, figure 9-6, if available. Otherwise, pop the flywheel loose by removing the flywheel nut and delivering a sharp hammer blow to a lead block held against the end of the crankshaft, figure 9-7. Another method is to back off the flywheel nut until it is about 1/3 off the crankshaft and then strike it sharply.

Do not hammer on the end of the crankshaft as this can damage the threads.

Adjust the breaker point gap and check the condenser and breaker point terminals for tightness, figure 9-8. Breaker points are set for 0.020 inch, in most cases, although the correct setting may vary according to the manufacturer. Turn the engine over until the breaker points reach their maximum opening, then check with a flat feeler gauge. Adjust the points if the setting is incorrect. Breaker points must line up, figure 9-9. Points can be cleaned if necessary. If the points are pitted or do not line up, figure 9-10, they should be replaced.

7. Fill the crankcase with clean oil of the correct type. (four-stroke cycle engines)

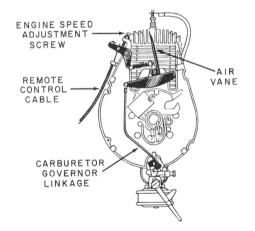

Fig. 9-5 Air vane governor parts

Fig. 9-7 Removing the flywheel with a soft hammer

Fig. 9-6 Removing a flywheel using a special puller

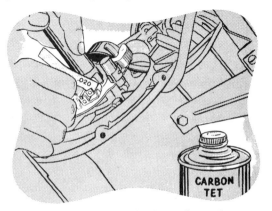

Fig. 9-8 Clean and check breaker points

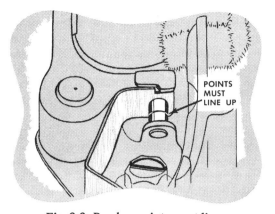

Fig. 9-9 Breaker points must line up.

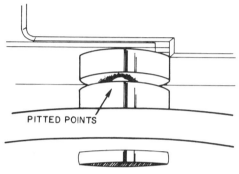

Fig. 9-10 Pitted breaker points should be replaced.

8. Fill the gasoline tank with regular gasoline. (four-stroke cycle engines)
Fill the gasoline tank with the correct mixture of gasoline and oil. (two-stroke cycle engines)

9. Start the engine.

10. Adjust the carburetor for peak performance. See carburetor parts shown in figure 9-11.

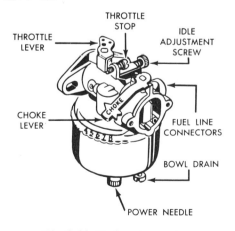

Fig. 9-11 Carburetor parts

RECONDITIONING

Reconditioning or overhauling an engine is generally the job for the mechanic—one who has the tools and know-how to do the job correctly. However, much reconditioning and overhauling can be done by the amateur who has the basic tools and the correct approach. Care and precision are very important. Every part must be in place correctly, with none missing or left over.

Before beginning the disassembly of an engine, provide a spot to put the parts as they are removed. One good method is to lay out a large sheet of paper and as the parts are placed on the paper, label them so they will not be lost. It is often difficult to remember just where a part came from when forty or fifty pieces are laid out, and it may be several days before the engine is reassembled.

A mechanic's handbook or service manual is essential for top-quality overhaul work. These books can sometimes be obtained directly from the engine manufacturer or borrowed from a mechanic. In only a few instances is this type of book supplied with the new engine. The service manual is quite detailed, fully explaining each step of overhaul or reconditioning. Allowances, clearances, torque data (see figure 9-12, page 166), and other specifications are also given. Without the use of the service manual there is too much guesswork involved. The amateur mechanic should have the service manual for the engine before any repair work is started.

Unless the engine owner is a skilled mechanic and has an adequately equipped repair shop, the owner cannot perform all types of engine repair and overhaul. Most owners would not find it practical to purchase the equipment necessary to do all types of overhaul and repair. Professional repairers may have invested from five hundred to several thousand dollars in tools and equipment:

CLINTON ENGINES TORGUE DATA — INCH POUNDS "Red Horse"		1600 A1600 A1690	1800 1890	2500 A2500	B2500 B2590 2790
Bearing Plate P.T.O.	Min. Max.	160 180	160 180	160 180	160 180
Back Plate to Block	Min. Max.	70 80	70 80	70 80	70 80

CLINTON ENGINES SERVICE CLEARANCES "Red Horse"		1600	A1600	A1690	1800	1890	2500	B2500	B2590	A2500	2790
Piston Skirt Clearance	Min. Max. Rework	.007 .009 .010	.007 .009 .010	.007 .009 .010	.007 .009 .010	.007 .009 .010	.0065 .0085 :010	.0065 .0085 .010	.0065 .0085 .010	.005 .007 .0085	.005 .007 .0085
Ring End Gap	Min. Max. Rework	.007 .017 .025	.007 .017 .025	.007 .017 .025	.007 .017 .025	.007 .017 .025	.010 .020 .028	.010 .020 .028	.010 .020 .028	.010 .020 .028	.010 .020 .028

CLINTON ENGINES TOLERANCES AND SPECIFICATIONS "Red Horse"	1600	A1600	A1690	1800	1890	2500	B2500	B2590	A2500	2790
Cylinder - Bore	2.8125 2.8135	2.8125 2.8135	2.8125 2.8135	2.9995 3.0005	2.9995 3.0005	3.1245 3.1255	3.1245 3.1255	3.1245 3.1255	3.1245 3.1255	3.1245 3.1255
Skirt - Diameter	2.8045 2.8055	2.8045 2.8055	2.8045 2.8055	2.9915 2.9925	2.9915 2.9925	3.117 3.118	3.117 3.118	3.117 3.118	3.1185 3.1195	3.1185 3.1195

Fig. 9-12 Torque data, clearances, tolerances, and specifications as excerpted from a service manual

magneto testers, valve seat resurfacers, air compressors, steam cleaners, special sharpeners, grinders, lapping stands, special factory tools, general repair tools, repair parts, and others.

However, a small investment in tools allows a person to perform many repair jobs. Most home workshops have some of the basic tools that are necessary. Several specialized tools can be added, figure 9-13. A tentative list of necessary tools includes the following:

- Screwdriver (various sizes)
- Combination wrenches (set)
- Adjustable wrench
- Socket set
- Torque wrench
- Deep well spark plug socket
- Needle nose pliers
- Feeler gauge
- Spark gap gauge
- Piston ring compressor
- Piston ring expander
- Valve spring compressor

Additional tools:

- Flywheel puller
- Compression gauge
- Tachometer
- Valve grinder—hand operated
- Arbor press

Reconditioning or overhaul of an engine involves four basic steps: disassembly, inspection of parts, repair or replacement of worn

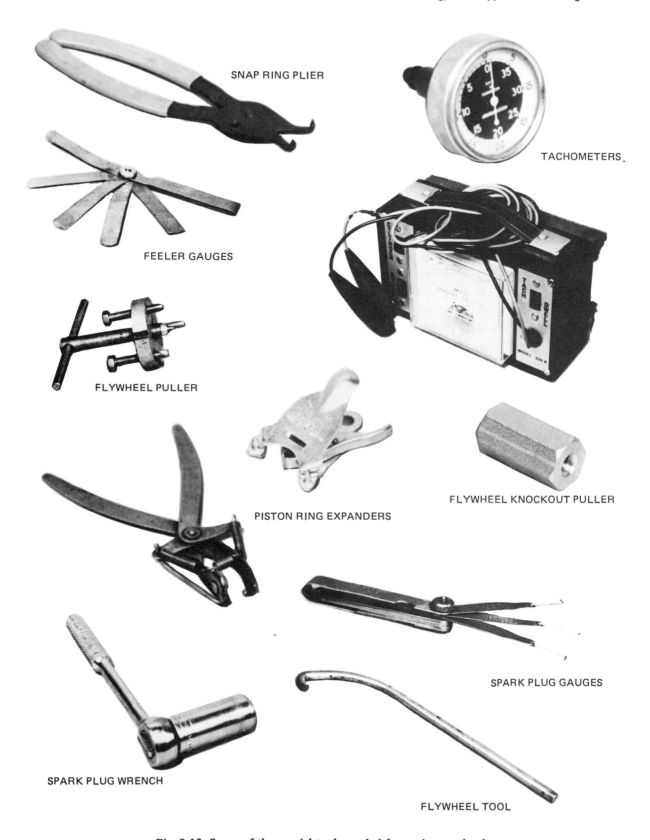

SNAP RING PLIER

TACHOMETERS

FEELER GAUGES

FLYWHEEL PULLER

PISTON RING EXPANDERS

FLYWHEEL KNOCKOUT PULLER

SPARK PLUG GAUGES

SPARK PLUG WRENCH

FLYWHEEL TOOL

Fig. 9-13 Some of the special tools needed for engine overhaul

or broken parts, and reassembly. The following disassembly procedure is a general guide for a four-stroke cycle engine.

1. Disconnect the spark plug lead and remove the spark plug.

2. Drain the fuel system—tank, lines, carburetor.

3. Drain the oil from the crankcase.

4. Remove the air cleaner.

5. Remove the carburetor.

6. Remove the metal air shrouding, gas tank, and recoil starter.

7. Remove the flywheel.

8. Remove the breaker assembly and push rod.

9. Remove the magneto plate assembly.

10. Remove the breather plate assembly (valve spring cover).

11. Remove the cylinder head.

12. Remove the valves.

13. Remove the base.

14. Remove the piston assembly.

15. Remove the crankshaft.

16. Remove the camshaft and tappets.

17. Remove the mechanical governor.

In the following discussion, disassembly, inspection, repair, and reassembly are discussed for each engine part. In actual practice a service manual would be followed for these steps. However, this information is an excellent source for the general procedure followed by most manufacturers. (1) Disconnect the spark plug lead and remove the spark plug, (2) drain the fuel system—tank, lines, carburetor, (3) drain the oil from the crankcase, (4) remove the air cleaner, (5) remove the carburetor, and (6) remove the metal air shrouding. Gas tank and recoil starter have been discussed in earlier units. Be certain to inspect

any parts removed for damage and wear. Replacement or repair may be indicated.

Removal of the flywheel is necessary to gain access to the ignition system parts. If a flywheel puller is available, it is best; however, the flywheel can be removed by striking the end of the crankshaft with a plastic or soft hammer. Take care not to damage the end of the crankshaft. The key on the crankshaft should not be lost. Keep it on the crankshaft or in a safe place. When reassembling the flywheel onto the crankshaft, carefully fit the key into the keyway (see figure 9-14).

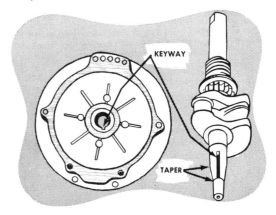

Fig. 9-14 Keyway on tapered end of the crankshaft and the flywheel

MAGNETO PARTS

The magneto parts—high-tension coil, breaker points, condenser—may need to be removed for checking or replacement. The coil may be visually inspected for cracks and gouges in insulation, evidence of overheating, and the condition of the leads where they go into the coil. If an ignition coil tester is available, the coil can be checked for firing, leakage, secondary continuity, and primary continuity. This checking should be done using the procedure recommended by the manufacturers.

Breaker points can be visually inspected for pitting, alignment, and contact surface.

They have a chrome or silvery appearance when new, and become gray as they are used.

The condenser can be visually inspected for dents, terminal lead damage, and broken mounting clip. A condenser tester, figure 9-15, is used to check the condenser for capacity, leakage, and series resistance. Follow the text procedure suggested by the test equipment manufacturer.

When the magneto is reassembled, the laminated iron core must be close to the flywheel magnet but not touching it. Generally, the closer it is, the better, but it should not be close enough to rub. This air gap is specified by the manufacturer. Too large an air gap can cause faulty magneto operation.

ENGINE TIMING

Engine timing refers to the magneto timing to the piston: the position of the piston just as the breaker points start to open. Timing is set at the factory but it is possible for timing to cause engine trouble. If the breaker points open too late in the cycle, power is lost; if the breaker points open too early in the cycle, detonation can result. Improper engine timing can be caused by the crankshaft and camshaft gears being installed one tooth off, spark advance mechanism stuck, breaker points incorrectly set, magneto assembly plate loose or slipped, or rotor incorrectly positioned. These causes of trouble depend on the engine and the type of magneto.

The position of the piston for timing a magneto varies from engine to engine. The piston can be at top dead center (TDC) or slightly before the top dead center (BTDC). Check engine specifications for this information.

One common method of checking timing is to locate TDC, figure 9-16. This is the point where the piston does not seem to move as the flywheel is rotated. A dial indicator can be used for greater accuracy. With TDC

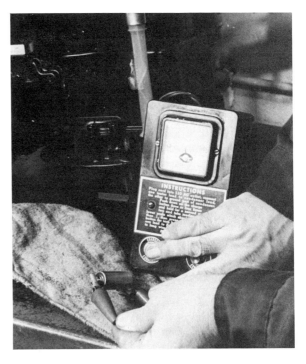

Fig. 9-15 Checking a condenser

located, place a reference mark on the flywheel and the magneto assembly plate. Now remove the flywheel. Rotate the crankshaft until the breaker points jsut open (0.001 inch). Carefully replace the flywheel, and put another mark on the magneto plate aligned with the flywheel reference mark. The difference between the marks on the magneto plate is the magneto timing to the piston. This can be figured in degrees by dividing the number of flywheel vanes into

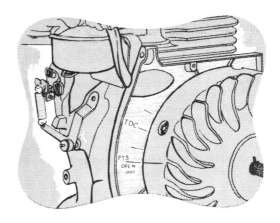

Fig. 9-16 Timing marks

360 degrees. (Twenty vanes; each vane 18 degrees.) The correct number of degrees of firing before top dead center is found in the engine specifications.

If the engine has a magnetic rotor magneto, the rotor and armature must be timed to the piston, figure 9-17. Timing is correctly set when the engine leaves the factory, but if the armature has been removed or the crankshaft or cam gear replaced, it is necessary to retime the rotor. Breaker points are first set correctly (0.020 inch). The rotor is on the shaft correctly and tightened. The armature is mounted but mounting screws are not tight. Turn the crankshaft until the breaker points just start to open. (Place a piece of tissue paper between the points to detect when the points let go.) Now turn the armature slightly until the timing marks on the rotor and the armature line up.

Two-cycle engine timing, figure 9-18, is quite similar. One common procedure is

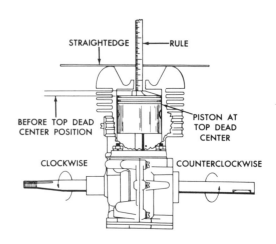

Fig. 9-18 Two-cycle engine timing

to remove the spark plug and, using a ruler and straightedge across the head, locate top dead center. With TDC located, back the piston down the cylinder the correct distance (check manufacturer's specifications for the measurement before TDC). Now the breaker points should just begin to open. If timing is incorrect, loosen the stator plate setscrew and rotate the stator plate slightly until the points just begin to open. Retighten the stator plate setscrew. See figure 9-19 for an example of two-cycle timing marks.

CYLINDER HEAD

The cylinder head must be removed if work is to be done on the piston, piston rings, connecting rod, or valves. It must also be removed if the combustion chamber is to be cleaned. The head screws or nuts should be removed and set aside. After the head is removed, the head gasket should be removed and discarded. Before reassembly, carbon deposits should be scraped off and the head cleaned. A new head gasket should be installed and the cylinder head screws or nuts tightened in the correct sequence. To be certain of the correct degree of tightness a torque wrench, figure 9-20, should be used (generally 14 to 18 ft.-lbs.).

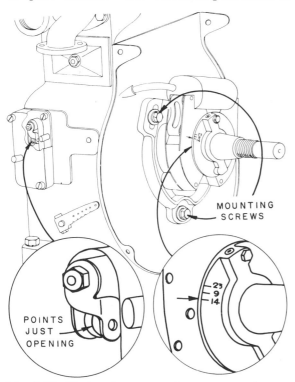

Fig. 9-17 Timing the rotor and armature to the piston. The numbers represent model numbers.

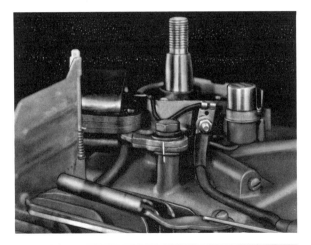

Fig. 9-19 Two-cycle timing marks

Fig. 9-20 Tightening the cylinder head bolts with a torque wrench

VALVES

The valves can easily be inspected. Inspection may show that they are stuck, burned, cracked, or fouled with carbon. Also, valve stems and valve guides may be worn or the valves and valve seats may need to be reground. Possibly the tappet clearance is wrong. Engine owners may be able to do some valve work themselves. However, a large amount of work or a complete renewal of the valve system should be done by a mechanic who has the tools and necessary experience. A complete valve job might include:

- Installing new valves
- Installing new valve guides
- Installing new valve seats
- Installing new valve springs
- Grinding valve seats
- Lapping valves

One of the first check points is the tappet clearance (space between tappet and end of valve). This clearance is checked with a feeler gauge, figure 9-21. On most small

Fig. 9-21 Checking tappet clearance

engines the clearance can be enlarged by grinding a small amount from the end of the valve. If the clearance is already too large, the valve would have to be replaced. Some small engines do have adjustable tappet clearances.

To remove the valves, first compress the valve spring, then slip off or slip out the valve spring retainers, sometimes called keepers. Pull the valve out of the engine. The valve spring and associated parts will also come out, figure 9-22.

With the valve out, the valve stem, face, and head can be closely inspected and cleaned. The valve guide and valve seat can also be inspected.

Many engines have replaceable valve guides. If these are worn or otherwise damaged, they must be pressed out using an arbor press, figure 9-23, or carefully driven out with a special punch. Also, the exhaust valve seat is removable on many engines; if the seat is beyond regrinding, remove it and reinstall a new valve seat. A special valve seat extracting tool, figure 9-24, is used for this job.

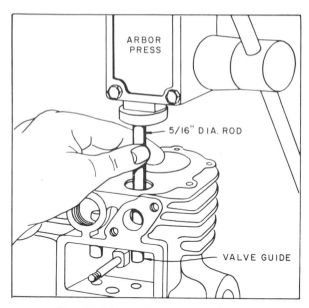

Fig. 9-23 Removing a worn valve guide

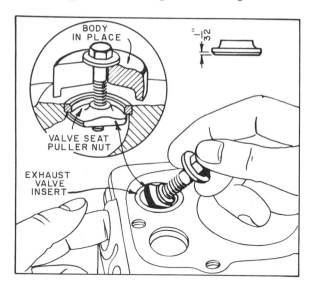

Fig. 9-24 Removing the exhaust valve seat with a special puller

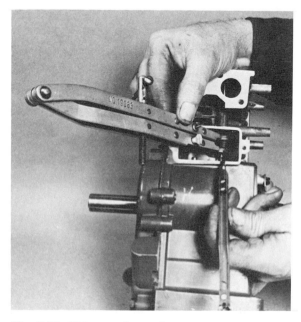

Fig. 9-22 Using a valve spring compressor to remove the valve springs

If the valves and/or valve seats need regrinding, this can be done with special valve grinding equipment, figure 9-25. However, in some cases hand valve grinders can be used, figure 9-26.

If either or both of these parts have been replaced or reground, the parts must be lapped, figure 9-27, to provide the perfect seal necessary for valve operation. When

Fig. 9-25 Valve grinder

Fig. 9-26 Reconditioning a valve seat

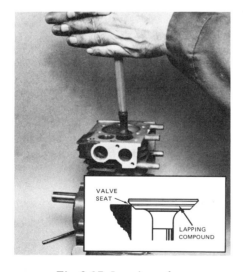

Fig. 9-27 Lapping valves

lapping, use a small amount of lapping compound. Rotate the valve against the seat a few times until the compound produces a dull finish on the valve face. Do not lap the valves too heavily. When the valves are replaced, oil the stems and be certain that the exhaust valve goes in the exhaust side and that the intake valve goes in the intake side.

Remove the engine from the base or sump, figure 9-28. This is done by loosening the bolts and breaking the seal. In most cases, a new gasket has to be installed upon reassembly.

To remove the piston assembly, remove the connecting rod cap and push the piston assembly up out of the engine. Mark the piston so it can be reinstalled the same way it comes out. The cylinder should be checked for score marks, figure 9-29, page 174. Scoring in the area of ring travel causes increased oil consumption and reduced engine power. Also the cylinder size should be checked with a cylinder gauge. If the cylinder appears to be in good condition, it can be deglazed with a

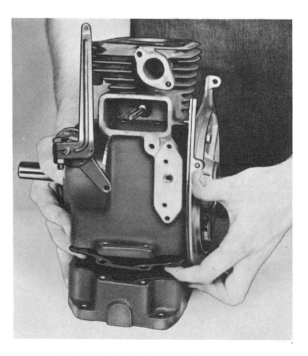

Fig. 9-28 Engine base, gasket, engine

finish hone, figure 9-30, to prepare the cylinder for new rings.

If the piston size and condition have checked out all right, the piston rings should be checked next. Check the edge gap with a feeler gauge, with the rings still on the piston. The correct edge gap clearance is found in the engine overhaul specifications. Remove the piston rings from the piston with a piston ring expander, figure 9-31. Carefully put the ring in the cylinder and check the end gap with a feeler gauge, figures 9-32 and 9-33. Too little end gap can cause the ring to freeze when it becomes hot and expands. Too much end gap may allow blow-by and a resulting loss of power. Piston ring grooves should be cleaned to remove any carbon

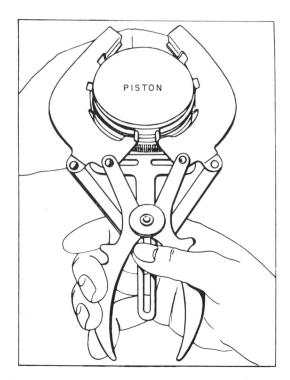

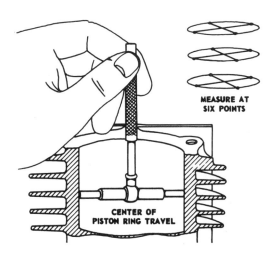

Fig. 9-29 Check cylinder bore

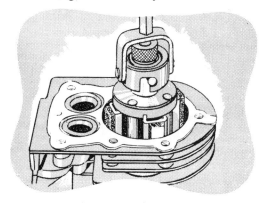

Fig. 9-30 Deglazing a cylinder with a finish hone

Fig. 9-31 Piston ring tool

accumulations. New piston rings are usually installed during engine overhaul.

Use a piston ring compressor, figure 9-34, to reinstall the piston assembly. Place the piston in the same way as it came out. Also, put the connecting rod cap on the same way as it came off, finding the match marks, fiure 9-35. A torque wrench should be used to tighten the connecting rod cap, figure 9-36. The crankshaft should be removed and checked for scoring and any metallic pickup. The journal and crank pin should be checked with a micrometer for roundness. The gear and keyway should be checked for wear.

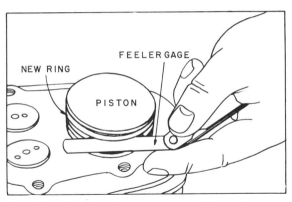

Fig. 9-32 Checking the ring groove with a feeler gauge

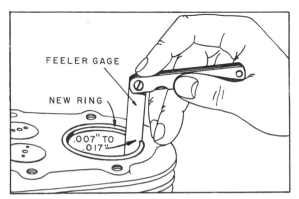

Fig. 9-33 Checking the ring gap with a feeler gauge (End gap)

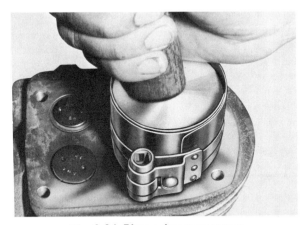

Fig. 9-34 Piston ring compressor

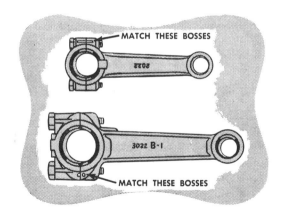

Fig. 9-35 Match boss marks when reassembling the connecting rod and connecting rod cap.

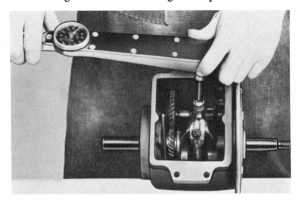

Fig. 9-36 Torque wrench used to correctly tighten connecting rod cap

Fig. 9-37 Pressing in the main ball bearing

In some cases the main ball bearings may come out with the crankshaft. Upon reinstallation, the bearings may have to be pressed into place. An arbor press is good for this job, figure 9-37.

The camshaft and valve tappets can also be removed for inspection and repair.

The camshaft pin should be driven out with a drift punch from the power takeoff side of the engine. Upon reassembly of the camshaft and crankshaft, figures 9-38 and 9-39, page 176, be certain to line up the timing marks.

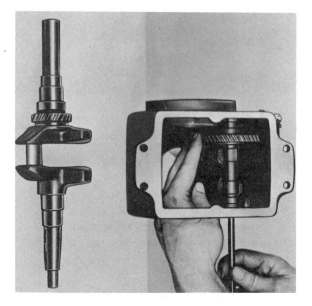

Fig. 9-38 Reinstalling the camshaft

Fig. 9-39 Reinstalling the crankshaft

REVIEW QUESTIONS

1. Define troubleshooting.

2. Of what value is a troubleshooting chart?

3. Can troubleshooting be done by an individual engine owner or operator?

4. Define engine tune-up.

5. Define engine reconditioning or overhaul.

6. Can any reconditioning or overhaul be accomplished by the amateur mechanic?

7. Why is the careful layout of parts important during disassembly?

8. Why is a mechanic's handbook or service manual essential for the reconditioning of an engine?

9. Explain engine timing.

10. What might a complete valve job include?

CLASS DISCUSSION TOPICS

- Discuss the troubleshooting chart.
- Discuss the steps in engine tune-up.
- Discuss the importance of care, accuracy, etc. in engine reconditioning and overhaul work.

- Discuss the tools necessary for reconditioning work.
- Discuss the steps in engine reconditioning.

CLASS DEMONSTRATION TOPICS

- Demonstrate troubleshooting by putting troubles into an engine and having students observe engine operation and then troubleshoot the engine. Simple examples: no fuel in tank, poor compression, shorting bar on plug, fuel shutoff valve closed, spark plug lead loose, bad spark plug, etc.
- Demonstrate how to tune up an engine.
- Demonstrate the correct use of the basic tools.
- Demonstrate timing an engine.
- Demonstrate the tightening sequence and correct torque on a cylinder head.
- Demonstrate removing and inspecting valves for wear and damage.
- Demonstrate checking tappet clearance.
- Demonstrate lapping valves.
- Demonstrate removing and inspecting piston assembly for damage and wear.
- Demonstrate removal and inspection of the camshaft and tappets.

LABORATORY EXPERIENCE 9-1
CLEAN AND ADJUST THE BREAKER POINTS

The breaker points are cleaned and reset. Students are to record their work after each step of the procedure is completed.

Procedure

Instructors may want to supplement or revise specific steps of this procedure since there are many makes of engines. The following procedure is a general guide.

1. Remove the spark plug lead.
2. Remove air shroud, grass screens, etc. in the area of the magneto.
3. Remove the flywheel if the points are located under it.
4. Remove the breaker point cover.
5. Rotate the crankshaft until maximum point opening is attained.
6. Check the breaker point gap, using a flat feeler gauge, before cleaning and resetting.
7. Clean the points with carbon tetrachloride on a lint-free cloth.
8. Loosen the breaker point assembly lock screw.

Part	Disassembly (nuts, bolts, etc.)	Operation performed	Tool used

9. With the points at maximum opening, turn the breaker point adjusting screw to attain the correct setting. Check the setting with a flat feeler gauge.

10. Tighten the breaker point assembly lock screw.

11. Rotate the crankshaft several times.

12. Recheck the breaker point setting with a flat feeler gauge.

REVIEW QUESTIONS

Reset the breaker points, then answer the following questions.

1. What was the breaker point gap before adjustment?

2. What is the correct breaker point setting for the engine?

3. Were the breaker points under the flywheel? If so, could the breaker cam be seen?

4. Were the breaker points mounted outside the flywheel?

LABORATORY EXPERIENCE 9-2
TEST SPARK PLUG AND MAGNETO OUTPUT

Note: Trouble in the ignition system can stem from the spark plug or from the magneto itself. It is often helpful when troubleshooting to determine if the magneto is delivering high-tension voltage to the spark plug, and if the spark plug is delivering a spark. Poor ignition can result in: engine missing, poor performance under heavy load, and hard starting. If trouble exists, these two simple checks can help to establish whether or not the ignition system is the source of the difficulty.

Students are to perform two tests: (A) the magneto output, and (B) the output at the spark plug. Record work after each step of the procedure is completed.

Procedure

Instructors may want to supplement or revise specific steps of this procedure since there are many makes of engines. The following procedure is a general guide.

A. Test the strength of magneto output.

   1. Remove the high-tension lead from spark plug.

   2. Hold the high-tension lead about 1/8 inch from the spark plug base.

   3. Turn the engine over as if it were being started.

   4. If a good spark jumps to the plug, the magneto output is satisfactory.

B. Test the strength of the spark at the spark plug.

   1. Remove the high-tension lead from the spark plug.

   2. Remove the spark plug from the engine and replace the high-tension lead.

Part	Disassembly (nuts, bolts, etc.)	Operation performed	Tool used

3. Lay the spark plug on the engine so the plug base touches bare metal.

4. Turn the engine over as if it were being started.

5. If a good spark jumps at the electrodes, the spark plug is good.

   Note: This is not an absolute test, since it is more difficult for a spark plug to fire under compression. If the plug is questionable, do not hesitate to install a new one.

REVIEW QUESTIONS

Upon completion of the two tests, answer the following questions.

1. Is it possible for the magneto to be good and the spark plug to be bad?

2. Is it possible to get the correct spark at the plug if the magneto is bad?

3. Why is it important to tighten the spark plug securely upon reassembly?

4. Describe the sparks that were seen. (Were they fat blue, thin blue, or yellowish blue?)

5. If the spark plug fires correctly when it is out of the cylinder, will it necessarily fire correctly when it is under compression?

LABORATORY EXPERIENCE 9-3
ADJUST ENGINE TIMING

Note: It is recommended that the adjustment of engine timing be attempted only when the engine repair manual is available for reference. Specific instructions and procedures should be followed. In order to do the work, specific engine data is necessary.

Students are to correctly adjust the engine timing on their assigned engines, and record their work. Instructors will provide a manufacturer's book which has specific instructions and procedures to follow.

Part	Disassembly (nuts, bolts, etc.)	Operation performed	Tool used

REVIEW QUESTIONS

Upon completion of the assigned work, answer the following questions.

1. Explain how to locate top dead center (TDC).

2. Why are the breaker points timed to open at TDC or before TDC? Why not after TDC?

3. List the steps that were followed in adjusting the engine timing.

4. What problems can incorrect engine timing cause?

LABORATORY EXPERIENCE 9-4
REPLACE PISTON RINGS

The piston ring assembly is removed from the engine, and then the piston rings are removed. The condition of the piston is inspected, new rings are installed, and the piston assembly is reassembled in the engine. When students actually install the new rings, they should have access to the piston ring clearance specifications for their engine. These clearances are checked with a feeler gauge to determine whether or not they are within the tolerance limits. Students are to record their work after each step of the disassembly procedure is completed.

Disassembly Procedure

Instructors may want to supplement or revise specific steps of this procedure since there are many makes of engines. The following procedure is a general guide for four-stroke cycle engines.

1. Remove the necessary parts to expose the crankshaft.
2. Remove the cylinder head.
3. Remove the connecting rod cap. (Note markings on the cap so that it can be reassembled in the same manner.)
4. Push the piston assembly up and out of the engine. (Mark the piston so that it can be reinstalled the same way it came out.)

Part	Disassembly (nuts, bolts, etc.)	Operation performed	Tool used

5. Inspect the condition of the piston.

6. Check the edge gap with a feeler gauge. (Rings are still on the piston.)

7. Remove the piston rings with a piston ring expander.

8. Carefully put the ring in the cylinder and check the end gap with a feeler gauge.

9. Clean the piston ring grooves, removing any carbon accumulations.

10. Install new rings or reinstall old rings using a piston ring expander.

11. Push the piston assembly back into the cylinder with the aid of a piston ring compressor.

12. Reassemble piston assembly to crankshaft.

REVIEW QUESTIONS

Upon completion of the assigned work, answer the following questions.

1. What are the two types of piston rings and what is the main function of each?

2. Describe the condition of the piston itself.

3. What were the edge gaps? Were they within tolerances?

4. What were the end gaps? Were they within tolerances?

5. Describe the markings or indications on the connecting rod and connecting rod cap that indicate correct reassembly.

6. What are some common troubles that are caused by bad piston rings?

LABORATORY EXPERIENCE 9-5
LAP VALVES

The valves are removed from the engine and inspected. (Valve seats are also checked.) Necessary parts are replaced, then the valves are lapped and reinstalled.

Note: If students are working on an instructional engine, the teacher may want them to assume that new parts have been installed even though they have not. The importance at this time is lapping the valves for practice and experience. Students are to record work after each step of the disassembly procedure is completed.

Disassembly Procedure

Instructors may want to supplement or revise specific steps of this procedure since there are many kinds of engines. The following procedure is a general guide.

1. Remove the valve spring cover exposing the valve springs.

2. Remove the cylinder head.

3. Compress the valve spring with a valve spring compressor.

4. Remove the valve spring retainers by slipping or flipping them out.

5. Remove the valve spring, still compressed.

6. Pull the valve out of the engine.

7. Inspect the valve and valve seat.

8. Replace the valve and valve seat, or regrind the valve and valve seat only upon the instructor's request.

Part	Disassembly (nuts, bolts, etc.)	Operation performed	Tool used

9. Lap the valves. Place a small amount of lapping compound on the valve face and rotate the valve against its seat a few times. Lap the valves until there is a thin, dull ring around the entire valve face. The ring indicates that the valve will seat well. Do not lap the valves more than is necessary.

10. Clean the valve and oil the valve stem.

Reassembly Procedure

Reverse the disassembly procedure. Remember to put the exhaust valve in the exhaust side and the intake valve in the intake side. On some engines a magnetic valve retainer inserter is a great help. The retainers often come out easily but are difficult to reinstall.

REVIEW QUESTIONS

Upon completion of the assigned work, answer the following questions.

1. Describe the condition of the valves.

2. Describe the condition of the valve seats.

3. What is the purpose of lapping valves?

4. What type of trouble can bad valves cause?

5. How is the exhaust valve different from the intake valve? Are there any special markings?

6. Can the exhaust valve seat on the engine be replaced?

7. Can the intake valve seat on the engine be replaced?

LABORATORY EXPERIENCE 9-6
CHECK ENGINE COMPRESSION

Note: Good compression is essential for top engine performance. Fuel mixture must be tightly compressed to ensure proper ignition and maximum power. Poor compression can be caused by worn piston rings, bad valves, worn or warped cylinders, leakage through the head gasket, or leakage around the spark plug. If an engine starts with difficulty, lacks power, or is sluggish, poor compression could be the cause. Checking an engine's compression is a part of most tune-up procedures. Engine manufacturers each have their own recommended methods for checking compression.

A compression check is performed on an engine. Students are to record their work after each step of the procedure is completed.

Procedure

Instructors may want to supplement or revise specific steps of these procedures since there are many makes of engines. The following procedures are general guides.

A. Checking compression without a compression gauge.

1. Remove the spark plug high-tension lead from the spark plug.

2. Turn the engine over slowly by hand. As the piston reaches top dead center, much resistance should be felt. As top dead center is passed, the piston should snap back down the cylinder, indicating good compression.

B. Checking compression with a compression gauge.

1. Remove the spark plug lead from the spark plug.

2. Remove the spark plug.

3. Carefully clean any dirt or foreign matter from around the spark plug hole.

4. Hold the compression gauge tightly against the spark plug hole. (Some gauges screw into the hole.)

Part	Disassembly (nuts, bolts, etc.)	Operation performed	Tool used

5. Turn the engine over as if starting it.

6. Read the compression gauge. Readings of 60 to 80 pounds per square inch (psi) generally indicate good compression; however, some engines may have a compression range of 110 to 120 psi. The exact data for the engine you are working on should be at hand.

REVIEW QUESTIONS

Upon completion of the compression check, answer the following questions.

1. Name some of the engine parts whose failure can cause poor compression.

2. What are the common troubles that are caused by poor compression?

3. Does correcting a condition of poor compression always require a major repair job or overhaul of the engine?

4. Name the manufacturer of the engine used in this experience. What type of compression check did the manufacturer recommend?

5. If a compression gauge was used, record the compression (in psi). Was this pressure within the manufacturer's acceptable range?

LABORATORY EXPERIENCE 9-7
TOLERANCES AND ENGINE MEASUREMENTS

Note: Making the following measurements is simple; only a feeler gauge is necessary. For tightening machine bolts and nuts a torque wrench is necessary. Of course, if instruments such as inside and outside micrometers, and dial indicators are available, the unit can easily be expanded.

The engine is disassembled so that certain clearances can be measured. Do not disassemble the engine any more than is necessary. Upon reassembly of the engine, nuts and machine bolts will be tightened to the proper degree with a torque wrench. If possible, obtain a repair manual for the engine and check the clearances against the tolerances. Also, use the torque data for the particular engine if it is available. Students are to record their work after each step of the disassembly procedure is completed.

General Procedure

Instructors may want to supplement or revise specific steps of this procedure since there are many makes of engines. The following procedure is a general guide. Read the entire procedure carefully before beginning any work.

Disassembly Procedure

Disassemble the engine. Instructors will give the necessary steps in the engine disassembly procedure.

Checking Procedure

Using a feeler gauge,

1. Check the camshaft end clearance.
2. Check the crankshaft end clearance.

Part	Disassembly (nuts, bolts, etc.)	Operation performed	Tool used

3. Check the connecting rod–large end side clearance.

4. Check the valve clearance for both intake and exhaust valves (measured between tappet and valve stem).

5. Check the piston skirt clearance at thrust face.

6. Check the ring end gap clearance.

7. Check the ring edge gap clearance.

8. Check the breaker points at maximum opening.

Reassembly Procedure

Using a torque wrench,

1. Tighten the connecting rod cap screws to 140-in.-lbs.

2. Tighten the cylinder head cap screw to 200 in.-lbs.

3. Tighten the flywheel nut to 45 ft.-lbs.

4. Tighten the spark plug to 27 ft.-lbs.

Note: Use the torque data for the particular engine if it is available.

REVIEW QUESTIONS

1. Why is precision workmanship necessary in servicing an engine?

2. Briefly define engine tolerance.

3. Briefly explain the use of the feeler gauge.

4. How does a torque wrench differ from other wrenches used by the mechanic?

5. Why is a torque wrench used instead of a regular wrench in tightening the various parts?

6. Explain the term ft.-lb.; in.-lb.

LABORATORY EXPERIENCE 9-8
TUNE-UP FOR SMALL ENGINES

The general steps in engine tune-up are performed on an assigned engine. Work carefully, bringing the engine up to its peak of cleanliness and operating efficiency. Students are to record their work after each step of the procedure is completed.

Procedure

Instructors may want to supplement or revise specific steps of this procedure since there are many makes of engines. The following procedure is a general guide.

1. Inspect the air cleaner, then clean and reassemble it.

2. Clean the gas tank, fuel lines, and any fuel filters or screens.

3. Check the compression.

4. Check the spark plug: clean, regap, or replace.

5. Check the operation of the governor.

6. Check the magneto.

7. Fill the crankcase with clean oil of the correct type.

8. Fill the gasoline tank with regular gasoline, being sure to mix oil with the gasoline if it is a two-stroke cycle engine.

9. Start the engine.

10. Adjust the carburetor for peak performance.

Part	Disassembly (nuts, bolts, etc.)	Operation performed	Tool used

REVIEW QUESTIONS

Upon completion of the assigned work, answer the following questions.

1. Explain the reason for engine tune-up.

2. Can engine tune-up be done by an engine owner at home?

3. If the engine fails to pass the compression check, what could be causing the trouble?

4. Explain the danger involved in operating a gasoline engine in a closed building.

5. Why is the spark plug on a single-cylinder engine of particular importance in engine tune-up?

6. List the tools that are used in tuning the engine.

UNIT 10 HORSEPOWER, SPECIFICATIONS, AND BUYING CONSIDERATIONS

OBJECTIVES

After completing this unit the student will be able to:

- discuss four kinds of horsepower.

- explain the terms engine torque, volumetric efficiency, compression ratio, and piston displacement.

- list at least five things to consider when buying an engine.

HORSEPOWER

Horsepower (HP) is the yardstick of the engine's power, its capacity to do work. James Watt, the inventor of the steam engine, devised the unit. He assumed that the average horse could raise 33,000 pounds one foot in one minute. An engine that can lift 33,000 pounds one foot in one minute is a one-horsepower engine. Of course, the engine can lift 8,250 pounds four feet in one minute, and still be delivering 1 HP.

As discussed in unit 1, the formula for horsepower is:

$$HP = \frac{Work}{Time \text{ (in minutes)} \text{ x } 33,000}$$

Remember that work is the energy required to move a weight through a distance. For example: lifting one pound, one foot, is one foot-pound of work. Lifting 50 pounds, two feet, is 100 foot-pounds of work. The element of time is not a factor.

In horsepower the time factor is added. For example, a boat with an outboard motor might cross a river in ten minutes. An identical boat with a larger motor might make the crossing in five minutes. The same amount of work has been accomplished by each engine but the larger horsepower engine did the work faster.

There are a variety of terms related to horsepower, such as brake horsepower, frictional horsepower, indicated horsepower, SAE or taxable horsepower, rated horsepower, developed horsepower, maximum horsepower, continuous horsepower, corrected horsepower, observed horsepower, and others. Several of these are discussed in the following paragraphs.

Brake horsepower is usually used by manufacturers to advertise their engine's power. Brake horsepower is measured either with a prony brake or with a dynamometer. The *prony brake* consists basically of a flywheel pully, adjustable brake band, lever, and scale measuring device. With the engine operating, the brake band is tightened on the flywheel and the pressure or torque is transmitted to the scale. The readings on the scale and other data are used to calculate the horsepower.

The *dynamometer* is a more recent device for measuring horsepower. The electric dynamometer contains a dynamo and as the engine drives the dynamo the current output can be carefully recorded. The more powerful the engine, the more current is produced.

Frictional horsepower is the power that is used to overcome the friction in the engine itself. The parts themselves absorb a certain amount of power. Losses included are mainly due to the friction of the piston moving in the cylinder but also included are losses due to oil and coolant pumps, valves, fans, ignition parts, and other accessories. These losses can amount to 10 percent of the brake horsepower.

Indicated horsepower is the power that is actually produced by the burning gases within the engine. It does not take into account the power that is absorbed by or used to move the engine parts. It is the sum of the brake horsepower plus the power used to drive the engine (frictional horsepower).

SAE or taxable horsepower is the horsepower used to compute the license fee for automobiles in some states.

$$\text{SAE HP} = \frac{D^2 \times \text{No. of Cylinders}}{2.5}$$

D = Diameter of bore in inches

Example: $\dfrac{(3.5 \times 3.5) \times 6}{2.5} = 29.4 \text{ HP}$

ESTIMATING HORSEPOWER

The following formula can be used to estimate the maximum horsepower of a four-stroke cycle engine.

$$\text{HP} = \frac{D \times N \times S \times \text{RPM}}{11,000}$$

D (Bore in Inches)

N (Number of Cylinders)

S (Stroke in Inches)

11,000 (Experimental Constant)

For a two-cycle engine, the formula can be used with 9000 as the experimental constant.

Engine torque is a factor that relates to horsepower. Torque is the twisting force of the engine's crankshaft. Torque can be compared to the force a person uses to tighten a nut.

At first a small amount of torque is used, but more and more torque is applied as the nut tightens; even when the nut stops turning, the person still may be applying torque. Motion is not necessary to have torque. Torque is measured in foot-pounds or inch-pounds. On an engine, the maximum torque is developed at speeds below the maximum engine speed. At top speeds, frictional horsepower is greater and volumetric efficiency is less.

Volumetric efficiency relates to horsepower also. It refers to the engine's ability to take in a full charge of fuel mixture in the short time allowed for intake. Engine design largely determines the volumetric efficiency of the engine. At high speeds the volumetric efficiency drops off. It can be increased with superchargers and turbochargers that force or blow air into the intake manifold. Also, volumetric efficiency can be increased by using multiple-barrel carburetors (2 barrel and 4 barrel) with the extra barrels opening up at high speed when the demand for air is greater.

Compression ratio is another factor that affects horsepower. This is the relationship of the volume of the cylinder when the piston is at the bottom of its stroke compared to the volume of the cylinder (and combustion chamber) when the piston is at top dead center. For small gasoline engines it may be 6:1; for automobile engines, the compression ratio may be 8:1 or higher. The higher the compression ratio, the greater the horsepower delivered when the fuel mixture is ignited.

Piston displacement is the volume of air the pistons displace from the bottom of their stroke to top dead center. Piston displacement also relates to horsepower. Generally, the greater the piston displacement the greater the horsepower. Most engines deliver 1/2 to 7/8 horsepower per cubic inch displacement. High performance engines develop about 1 horsepower per cubic inch displacement. Supercharged engines can deliver much more than 1 horsepower per cubic inch displacement. Displacement equals *area of bore x stroke x number of cylinders,* and is expressed in cubic inches.

SPECIFICATIONS

Specifications for several models of some of the leading manufacturers of engines are listed below. In considering the best engine for a certain job, such information should be studied.

CUSHMAN MOTORS, Lincoln, Nebraska — Principal Usage: Utility Vehicles*

Model*	Rated HP	RPM	Bore	Stroke	Disp. Cu. in.	Comp. Ratio	2 or 4 Cycle	Weight
100	9		3.5	2.25	21.58	6.85/1	4	
200	18		3.5	2.25	43.16	6.85/1	4	
*Partial List of Motors								

KOHLER CO., Kohler, Wisconsin — Principal Usage: All engine powered equipment

Model*	Rated HP	RPM	Bore	Stroke	Disp. Cu. in.	Comp. Ratio	2 or 4 Cycle	Weight
K91	4.1	4000	2 3/8	2	8.86	6.5/1	4	41
K161	7.0	3600	2 7/8	2 1/2	16.22	6.25/1	4	65
KV161	7.0	3600	2 7/8	2 1/2	16.22	6.25/1	4	65
L160	6.5	3600	2 7/8	2 1/2	16.22	6.25/1	4	106
K241	9.5	3600	3 1/4	2 7/8	23.9	6.00/1	4	105
K331	12.5	3200	3 5/8	3 1/4	33.6	6.25/1	4	173
K662	24.0	3200	3 5/8	3 1/4	67.2	6.00/1	4	250
*Other models too numerous to mention								

JACOBSEN MANUFACTURING CO., Racine, Wisconsin — Principal Usage: Lawn Equipment

Model	Rated HP	RPM	Bore	Stroke	Disp. Cu. in.	Comp. Ratio	2 or 4 Cycle	Weight
J-125-H	2.25		2.0	1.5	4.71	5.5/1	2	
J-100-H	1.8		2.0	1.5	4.71	5.5/1	2	
J-125-V	2.25		2.0	1.5	4.71	5.5/1	2	
J-175-H	3.0		2.12	1.75	6.2	5.5/1	2	
J-175-V								
J-225-V	4.0		2.25	2.0	7.95	5.3/1	2	
J-321-V	3.0		2.12	1.75	6.2	5.0/1	2	
J-321-H								

BRIGGS AND STRATTON CORP., Milwaukee, Wisconsin — Principal Usage: General Power Use

Model*	Rated HP	RPM	Bore	Stroke	Disp. Cu. In.	Comp. Ratio	2 or 4 Cycle	Weight
92500	3	3600	2 9/16	1 3/4	9.02		4	19.5
100900	4	3600	2 1/2	2 1/8	10.43		4	30.5
130900	5	3600	2 9/16	2 7/16	12.57		4	30.75
60100	2	3600	2 3/8	1 1/2	6.65		4	22.25
80100	2.5	3600	2 3/8	1 3/4	7.75		4	22.25
80300	3	3600	2 3/8	1 3/4	7.75		4	25.25
190400	8	3600	3	2 3/4	19.44		4	45

*Other models too numerous to mention

CLINTON ENGINES CORP., Maquoketa, Iowa — Principal Usage: General Power Use

Model*	Rated HP	RPM	Bore	Stroke	Disp. Cu. In.	Comp. Ratio	2 or 4 Cycle	Weight
A2100	2.25	3600	2 3/8	1 5/8	7.2		4	21 1/2
100	2.50	3600	2 3/8	1 5/8	7.2		4	23
4100	2.75	3600	2 3/8	1 7/8	8.3		4	21 1/2
3100	3.00	3600	2 3/8	1 7/8	8.3		4	23
V1000	3.25	3600	2 3/8	1 7/8	8.3		4	36
V1100	3.75	3600	2 3/8	2 1/8	9.5		4	36
B1290	4.00	3600	2 15/32	2 1/8	10.2		4	45
V1200	4.50	3600	2 15/32	2 1/8	10.2		4	40
A1600	6.30	3600	2 13/16	2 5/8	16.3		4	87
B2500	9.60	3600	3 1/8	3 1/4	25.0		4	103
2790	10.30	3600	3 1/8	3 1/4	25.0		4	103

*Other models too numerous to mention

LAWN BOY, Galesburg, Illinois — Principal Usage: Lawn Mowers

Model*	Rated HP	RPM	Bore	Stroke	Disp. Cu. In.	Comp. Ratio	2 or 4 Cycle	Weight
C-10		4000	1 15/16	1 1/2	4.43	6.5/1	2	
C-12AA		4000	2 1/8	1 1/2	5.22	6.5/1	2	
C-18			2 3/8	1 1/2	6.65		2	
D-400			2 3/8	1 1/2	6.65		2	

*Other models too numerous to mention

WISCONSIN MOTOR CORP., Milwaukee, Wisconsin — Principal Usage: Heavy Duty Industrial

Model*	Rated HP	RPM	Bore	Stroke	Disp. Cu. In.	Comp. Ratio	2 or 4 Cycle	Weight
ACN	6	3600	2 5/8	2 3/4	14.88		4	76
BKN	7	3600	2 7/8	2 3/4	17.8		4	76
AENL	9.2	3600	3	3 1/4	23		4	110
AEH	7.4	3200	3	3 1/4	23		4	130
AGND	12.5	3200	3 1/2	4	38.5		4	180
THD	18	3200	3 1/4	3 1/4	53.9		4	220
VE4	21.5	2400	3	3 1/4	91.9		4	295
VF4	25	2400	3 1/4	3 1/4	107.7		4	295
VH4	30	2800	3 1/4	3 1/4	107.7		4	310
VG4D	37	2400	3 1/2	4	154		4	410
VR4D	56.5	2200	4 1/4	4 1/2	255		4	775

*Other models too numerous to mention

GRAVELY TRACTORS, INC., Dunbar, West Virginia — Principal Usage: Gravely Tractors

Model	Rated HP	RPM	Bore	Stroke	Disp. Cu. In.	Comp. Ratio	2 or 4 Cycle	Weight
L	6.6	2600	3 1/4	3 1/2	29.0	5/1	4	296

O & R ENGINES, Los Angeles, California — Principal Usage: General Power

Model	Rated HP	RPM	Bore	Stroke	Disp. Cu. In.	Comp. Ratio	2 or 4 Cycle	Weight
13B	1.0	6200	1.25	1.096	1.34	9/1	2	3.9
13A	1.0	6600	1.25	1.096	1.34	9/1	2	3.9
20A	1.6	7200	1.437	1.250	2		2	4.9

CHRYSLER OUTBOARD CORP. (Formerly West Bend) Hartford, Wisconsin
Principal Usage: Chain Saws, Scooters, Carts

Model	Rated HP	RPM	Bore	Stroke	Disp. Cu. In.	Comp. Ratio	2 or 4 Cycle	Weight
27824	3	4500	2	1 5/8	5.1	5.5/1	2	13 1/2
27825	3	4500	2	1 5/8	5.1	5.5/1	2	13 1/2
27852	3	4500	2	1 5/8	5.1	5.5/1	2	13 1/2
27854	3	4500	2	1 5/8	5.1	5.5/1	2	13 1/2
27612	5	5500	2 1/4	1 3/4	7.0	5.7/1	2	13 1/2
2760	5	5500	2 1/4	1 3/4	7.0	5.7/1	2	13 1/2

D. W. ONAN AND SONS, INC., Minneapolis, Minnesota
Principal Usage: Compressors, Truck Ref. Mowers, Go-Carts, Scooters, etc.
(Onan Generator units not listed)

Model	Rated HP	RPM	Bore	Stroke	Disp. Cu. In.	Comp. Ratio	2 or 4 Cycle	Weight
AS	5.5	3600	2 3/4	2 1/2	14.9	6.25/1	4	85
CCK	12.9	2700	3 1/4	3	50.0	5.50/1	4	148

TECUMSEH PRODUCTS CO., Grafton, Wisconsin – Principal Usage: General Power Use

Model*	Rated HP	RPM	Bore	Stroke	Disp. Cu. In.	Comp. Ratio	2 or 4 Cycle	Weight
AH47	3.2	4800	2	1 1/2	4.7		2	13 3/4
AH81	5.5	5000	2 1/2	1 5/8	7.98		2	13 3/4
AV47	2.2	3800	2	1 1/2	4.7		2	14 1/2
V55	5.5	3600	2 5/8	2 1/4	13.53		4	36 1/2
HR30	3.0	3600	2 5/16	1 13/16	7.61		4	28 1/4
HB30	3.0	3600	2 5/16	1 13/16	7.61		4	24

*Other models too numerous to mention

MCCULLOCH CORPORATION, Los Angeles, California – Principal Usage: Chain Saws

Model*	Rated HP	RPM	Bore	Stroke	Disp. Cu. In.	Comp. Ratio	2 or 4 Cycle	Weight
1-40		6000	2 1/8	1 3/8	4.9	5.5/1	2	18
1-50		6000	2 1/8	1 3/8	4.9	5.5/1	2	18
1-70		7000	2 1/4	1 1/2	5.3	7.0/1	2	21
1-80		7000	2 1/8	1 1/2	5.3	7.5/1	2	25

*Other models too numerous to mention

BUYING CONSIDERATIONS

The person who is shopping for a gasoline engine will have no trouble finding an engine for the job. There are many manufacturers of small gasoline engines and they produce engines in a variety of horsepower that are designed and engineered to satisfy every need. It is not uncommon to find a manufacturer's basic model with variations to adapt the engine for many different jobs. The following are several points to consider in buying an engine.

Cost. Beware of bargains; a good bargain can be found now and then but the general rule, "You get what you pay for," is a good one. This does not mean that the most expensive engine is the best buy. The engine must be the right one for the intended usage. For cutting the grass once a week, an economical, dependable engine should be enough. However, if the engine will be exposed to continuous heavy usage it would be wise to buy a heavy-duty engine, one made to operate for a long period of time without failure. The additional cost is worthwhile in this case.

Reputation of the Manufacturer. Be certain the manufacturer's past record in the field justifies faith in the new engines. Talk with engine dealers and individual owners of engines. Their opinions and experience can help in making the best selection.

Repair Parts. If an engine breaks down and requires a new part, will it be readily available? Can the part be found at a local dealer or will it be necessary to send to a factory that may be many miles away? Delays caused by breakdowns can be costly, as well as annoying. The availability of service and repair parts is an important consideration.

Power Requirements. Be certain the engine is big enough for the job. Constant overloading or constant operating at full throttle shortens engine life. A dealer can assist in finding the correct power plant for the job.

Manufacturers today produce engines that operate on either the four-cycle or the two-cycle principle; both have their place and both are entirely successful. Some engines are water cooled, some are air cooled; some have simple splash lubrication, some have pump lubrication; some have fuel pumps, and some do not. In fact, no two engines look exactly alike or operate exactly alike.

Engine Warranties. Most engine manufacturers give a warranty to the purchaser of a new engine. A common type is good for ninety days. If, during the ninety days after purchase, any parts fail due to defective material or workmanship, the manufacturer will repair or replace the defective part.

In most cases the part or engine must be returned to the factory or to an authorized distributor. Most manufacturers require the owner to pay shipping charges to the factory. In some cases, the owner must also pay any labor costs involved in the repair.

If the engine has been damaged through misuse, negligence, or accident, most warranties are void. A few manufacturers require that the engine be registered shortly after purchase; if the engine is not registered, the warranty is not valid. Generally, engine components such as magnetos, carburetors, starters, etc. are only covered by the terms of their individual manufacturer.

REVIEW QUESTIONS

1. If an engine can lift 150,000 pounds in three minutes, what is its horsepower?

2. In general, can larger horsepower engines do work faster than smaller horsepower engines?

3. What type of horsepower rating do most manufacturers use in advertising their engines?

4. What is a dynamometer?

5. In what way is SAE horsepower used?

6. Estimate the horsepower of a two-stroke cycle having a bore of 2 inches, one cylinder, a stroke of 1 1/2 inches, and operating at 3600 revolutions per minute.

7. What is engine torque?

8. What is volumetric efficiency?

9. What is the relationship between compression ratio and horsepower?

10. What is the relationship between piston displacement and horsepower?

11. List four important considerations for a person to think about before purchasing an engine.

CLASS DISCUSSION TOPICS

- Discuss how horsepower increases as RPM increases. (Remember that engines deliver much less than their rated horsepower at slow speeds.)

- Discuss and estimate the horsepower of a live engine in class. How does estimated horsepower compare with rated horsepower?

- Discuss how engine torque, volumetric efficiency, compression ratio, and piston displacement are functions of engine design.

- Discuss why there is a limit to the compression ratio that can practically be built into engines.

- Discuss the buying considerations for engines.

UNIT 11 ROTATING COMBUSTION ENGINES

OBJECTIVES

After completing this unit the student will be able to:

- prepare an RC engine for service.
- perform routine maintenance on an RC engine.
- prepare an RC engine for storage.

A new power-producing device, or prime mover, has become a member of the internal combustion engine family: the *rotating combustion (RC) engine,* also referred to as the Wankel engine. The key word is rotating (versus reciprocating). In an RC engine, heat energy rotates a three-lobe rotor (piston) within a chamber (cylinder) that has the shape of a fat figure eight. The motion of the rotor is in one direction only, figure 11-1, not back and forth as in a reciprocating engine.

For years, engineers have recognized the advantages of smooth rotary motion over reciprocating motion. In a reciprocating internal combustion engine, much power is lost in overcoming the inertia of the piston at both the top and the bottom of its stroke when the motion must be stopped and the

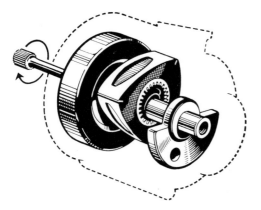

Fig. 11-1 The NSU-Wankel combustion engine utilizes rotary, rather than reciprocating, motion.

direction reversed. This power loss has resulted in the design of lightweight pistons and connecting rods in modern engines. Although a properly tuned reciprocating engine may seem smooth, there is a great deal of vibration as a result of its design.

Considering these facts, it may seem ususual that reciprocating engines are in wide use and rotating combustion engines are just coming into use. Scientists and engineers have thoroughly investigated both the rotating and reciprocating principles. However, since the reciprocating engine presented fewer manufacturing problems and a design that was easily understood, it was adopted. For about eighty years, work toward its perfection has been intensely continued throughout the industrialized nations of the world. The reciprocating principle was applied in many power devices: reciprocating steam engines, diesel engines, and reciprocating gasoline engines.

The development of steam turbines (and diesels) all but eliminated the reciprocating steam engine from the list of modern prime movers. In the commercial aircraft industry and in military aviation, the gas turbine or jet engine largely replaced the piston aircraft engine. Gas turbines are also being used in some applications for stationary and vehicular power plants. Rotating shaft turbines have made notable progress and today are firmly established as a major prime mover.

However, rotating combustion engines are not turbines that use hot gases directed against the vanes of a spinning wheel. Rotating combustion engines operate on the basic cycle of intake, compression, power, and exhaust, occurring in a repetitive pattern.

DEVELOPMENT AND APPLICATION OF ROTATING COMBUSTION ENGINES

From as early as 1799, inventive people experimented with different variations of the RC engine. As these inventors worked on rotating internal combustion engines, two problems persisted: the complex geometry (design) involved, and sealing the power or the pressures created by the burning and expanding gases. Sealing problems in the rotating engine were not as simple to solve as those presented by the round piston in a round cylinder of a reciprocating engine.

Felix Wankel, a German engineer, was attracted to the rotating combustion engine—its problems, and its potential usefulness. For years, he worked in the area of developing better sealing techniques. Gradually, his expertise became known, and he set up his own engineering and research laboratory. In 1951, Wankel made contact with the NSU research department (NSU is an engine manufacturer in Germany). This meeting eventually resulted in getting rotating combustion engines out of the laboratory and into industrial production. In 1958, the Curtiss-Wright Corporation, a company with long and broad experience in engine manufacture in the United States, became the licensee for the NSU/Wankel engine in North America. (See figures 11-2 and 11-3 for illustrations of the Wankel engine and its inventor.) Other prominent manufacturers have joined the ranks of RC license holders. Several of these companies are currently producing RC engines for the

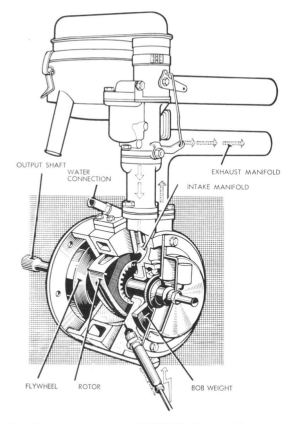

Fig. 11-2 Cross section of NSU-Wankel rotating combustion engine shows its efficient, yet simple, design.

Fig. 11-3 Felix Wankel with his rotating combustion engine

mass market, while others are still in their developmental, testing, and limited production phases. Some RC engine applications are shown in figure 11-4.

Curtiss-Wright has successfully manufactured and tested several versions of the Wankel engine. In the automotive field, NSU (now Audi NSU Auto Union AG) of Germany produced the NSU Spider from 1964 to 1967 and is currently producing the RO80 automobile. Toyo Kogyo introduced the Wankel engine on its Cosmo Sport auto of 1967 and

is now producing the Mazda R-100 and Mazda RX-2.

Manufacturers producing smaller horsepower RC engines are soon to become numerous. Fichtel and Sachs AG of West Germany is the largest producer, manufacturing RC engines which range from 6 to 20 HP for outboard motors, generators, pumps, lawn mowers, sailplanes, cycles, and snowmobiles. In the United States, their engines are probably most commonly seen on snowmobiles. Yanmar Diesel Co. of Japan is marketing

A. The twin rotor RO-80 Engine used on the latest NSU automobiles; B. The NSU RO-80 Engine adapted for marine application; C. Cessna Cardinal Airplane with Curtiss-Wright RC2-60 Engine. D. SACHS Wankel used to power a motorcycle; E. Yanmar Rotary Outboard Motor; F. Fichtel and Sachs Engine used to power a snowmobile.

Fig. 11-4 Some RC engine applications

20- and 45-horsepower RC outboard motors and has plans for a snow thrower, garden tractor, and other applications. Both Suzuki and Yamaha, Japanese manufacturers, are now developing Wankel-powered motorcycles. The Outboard Marine Corporation became the first United States firm to mass-produce an RC engine, introducing it in September, 1972, on their Johnson and Evinrude snowmobiles. The engine is capable of developing 35 HP at 5500 RPM and weighs only 62 pounds.

HOW THEY WORK

Like the reciprocating engine, the rotating combustion engine must bring fuel into the engine, compress the fuel, ignite the fuel, allow the pressure of the burning fuel to move the rotor (piston), and allow the burned gases to be exhausted from the engine. However, this is done in a rotating motion rather than a reciprocating motion.

The construction and design that makes this possible is a *rotor* (piston), which is a slightly rounded equilateral triangle rotating on an eccentric (off-center) portion of the main drive shaft, figure 11-5. The relationship of the eccentric bearing and the rotor radius generates the epitrochoidal shape of the housing chamber. The rotor apexes (or

tips) follow the shape of the chamber as the rotor is revolved about the shaft eccentric. Remaining in constant, tight contact with the chamber walls, the rotor apexes separate the open volume between the rotor sides and the chamber into three areas which are continually enlarging and reducing as the rotor turns. The housing is provided with intake ports, spark plug, and exhaust ports located at the proper positions to take advantage of the changing volume and provide for intake, compression, power, and exhaust.

OPERATING CYCLE

Figure 11-6, page 210, shows the operating cycle or phasing of the RC engine, which is similar to a three-cylinder piston engine since there are three sides to the rotor. This discussion of the operating principles follows the action of: side AC as it finishes its exhaust phase and takes fuel into the chamber; side AB as it compresses the fuel, ignites the fuel, and begins the power phase; and side BC as it completes the power phase and begins its exhaust phase.

Positions 1, 2, 3, and 4 (numerals circled in figure 11-6) show the uncovering of the intake port and the increasing volume of the chamber. The fuel mixture is pushed into the

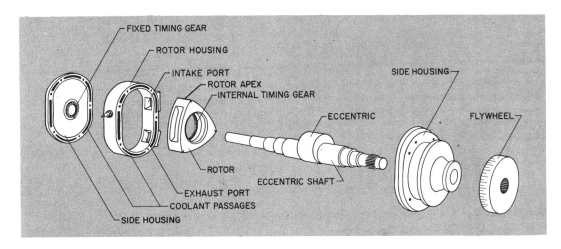

Fig. 11-5 Basic parts of the rotary engine

chamber by atmospheric pressure. The engine has a conventional carburetor and the operating principle is the same as that applied to reciprocating engines.

Positions 5, 6, and 7 in figure 11-6 show the changing shapes of the chamber during compression. As the rotor continues to turn, the fuel mixture is compressed into a smaller and smaller space. Compression ratios are designed to meet particular needs and they usually approximate those of similar sized reciprocating engines. For example, a compression ratio of 9 to 1 is used in the NSU RO 80 automobile engine. At position 7, maximum compression is attained and the spark plug fires, igniting the mixture. The spark plug tip is recessed in a hole so that the rotor can sweep by without hitting it.

Positions 8, 9, and 10 in figure 11-6 show the expanding gases exercising pressure against the face of the rotor. This is equivalent to the power stroke.

Positions 11, 12, and 1 make up the exhaust phase of the cycle. As the apex passes the exhaust port, the exhaust is forced from the engine.

The design of the engine permits the rotor to travel at one-third the speed of the main drive shaft. For instance, if an RC engine has a shaft speed of 6,000 RPM, the rotor travels at 2,000 RPM and there are 6,000 power strokes per minute. There is a power stroke for every revolution of the drive shaft. Speed ranges of 2,000–17,000 RPM have been reached. Most small RC engines have rated speeds of 4,500–5,500 RPM.

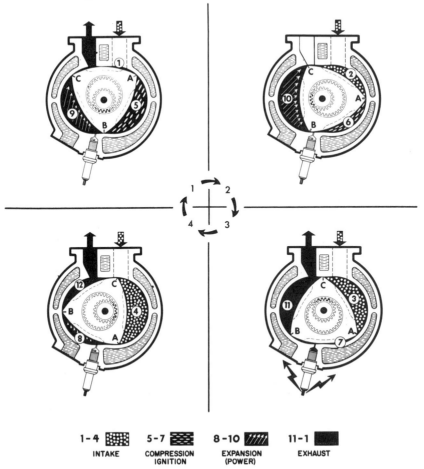

1-4	5-7	8-10	11-1
INTAKE	COMPRESSION IGNITION	EXPANSION (POWER)	EXHAUST

Fig. 11-6 Operating principles of the rotating combustion engine

CONSTRUCTION OF THE ROTATING COMBUSTION ENGINE

The different makes of rotating combustion engines share the same basic operating principles and parts; however, there are variations in engineering approaches, just as there are in reciprocating engines. The following description of the Curtiss-Wright RC-60, figure 11-7, is typical of RC engines.

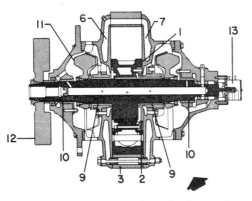

RC-60 — Longitudinal and Cross Sections

1. Rotor With Internal Rotor Gear
2. Stationary Gear
3. Rotor Housing
4. Exhaust Port
5. Spark Plug
6. Side Housing — Drive Side
7. Side Housing — Anti-Drive Side
8. Intake Port
9. Main Bearing (Inner)
10. Main Bearing (Outer)
11. Balance Weight
12. Flywheel
13. Ignition Contact Maker

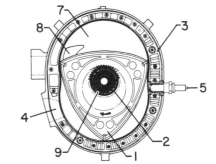

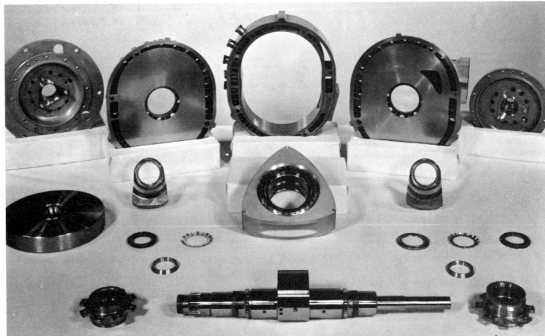

Fig. 11-7 The Curtiss-Wright experimental engine, RC-60

Other manufacturers' specifications are given where appropriate.

The rotor is a hollow casting of either aluminum or cast iron, figure 11-8. The three rotor flanks are exposed to a great concentration of heat; therefore, cooling oil is circulated through the rotor. Oil, under pressure from the oil pump, enters the rotor through drilled passages in the main shaft, figure 11-9, picks up heat, and then is discharged back into the system. A large hole machined in the center of the rotor accommodates the eccentric of the main shaft. A ring gear is secured to the rotor and meshes with a stationary gear on one of the end housings. These gears ensure that the correct relationship is followed by the engine parts as the rotor follows its path.

Devising effective seals for the rotor was accomplished only after much testing and development. The rotor must be sealed on the sides and also at the three rotor apexes, figure 11-10.

The seals, figure 11-11, are made of alloy cast iron which does not wear excessively nor does it cause excessive wear on the rotor housing. Light springs, figure 11-12, push the apex seals and side seals against the housing surfaces to provide an airtight seal. The apex seal may be traveling at the sliding velocity of up to 108 feet per second.

Fig. 11-8 Hollow construction of rotor

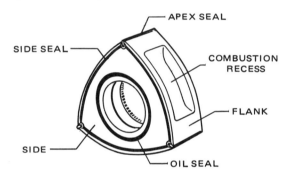

Fig. 11-10 Location of rotor seals

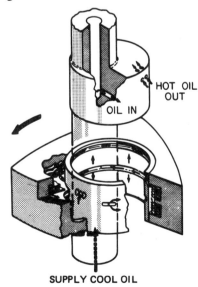

Fig. 11-9 Oil flow through main shaft, eccentric, and rotor

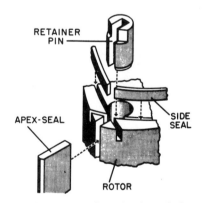

Fig. 11-11 Apex seals and side seals for rotor

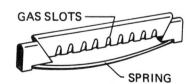

Fig. 11-12 Construction of apex seal used on NSU RO-80 engine

Lubrication in some form must be provided, just as piston rings must be lubricated. NSU/Wankel in Germany employs a separate oil supply delivered to the intake charge of fuel mixture. Curtiss-Wright provides lubrication by means of metering oil seal rings in the rotor itself. A rotating combustion engine consumes oil at about the same rate as a conventional reciprocating engine.

The forces that move the rotor are transmitted to the main shaft by means of an eccentric section on the shaft. The eccentric enables the rotor to follow the shape of the housing. The main shaft is held and centered by the two end housings. One end of the shaft contains the various gear takeoffs for the engine, such as the water pump, oil pump, and distributor shaft.

Rotating combustion engines are either liquid or air cooled. In the liquid-cooled RC engines, cooling water is circulated around the main chamber housing and around the end

units, figure 11-13. Special attention is given to the combustion area, figure 11-14. Combustion for each rotor flank always occurs at

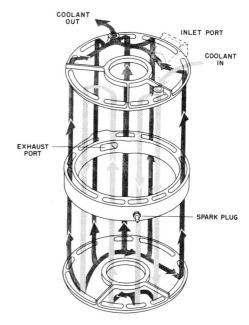

Fig. 11-13 Schematic of housing coolant flow

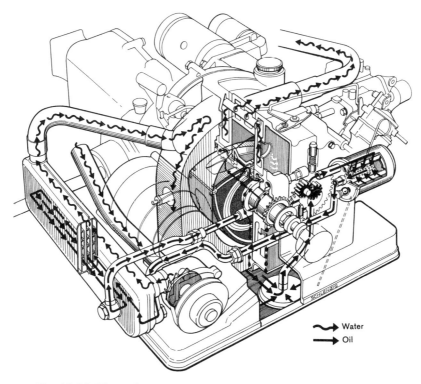

Fig. 11-14 Flow of water and oil coolant through NSU Spider engine

the same point, and enough coolant must be present at this area.

Air cooling, such as in the Fichtel and Sachs AG and the Outboard Marine Corporation (OMC) RC engines, is achieved by blowing cool air across ribs or fins of the housings, and trochoid housing, figure 11-15. The cool air is directed toward the fins by pressure fans or centrifugal blowers, figure 11-16. In areas of greatest heat concentration, the cooling ribs are more numerous, larger, and longer to provide better heat dissipation. Air-cooled RC engines have not experienced noticeable heat distortion problems. The rotor is cooled by the incoming fuel mixture which travels from the carburetor at the drive side, through

triangular windows at the rotor points, to the end side housing, before entering the intake ports around the rotor housing. This cooling technique is known as *charge cooling.*

One significant change that Curtiss-Wright has incorporated into their RC engines is the intake ports. The Curtiss-Wright engine has side intake ports, figure 11-17, that provide improved low-speed operation and lessen fuel consumption. The NSU/Wankel engine has intake ports located in the main housing. Both companies use exhaust ports in the rotor housing.

The OMC engine, figure 11-18, has two intake ports to meet the need for smooth idling and full-power operation. A port on the side housing provides fuel mixture for idling. An additional intake port located on the rotor housing opens to allow more airflow after the first 15 degrees of throttle movement. This power port provides for top high-speed performance, figure 11-19, page 216.

In the Fichtel and Sachs engines, lubricating oil is mixed with the fuel as in two-cycle engines. The ratio of oil to gasoline is 1 to 50 or 40 to 1. This method provides ample lubrication for the needle bearings or roller bearings and the side and apex seals of the

Fig. 11-15 Air-cooled RC engine

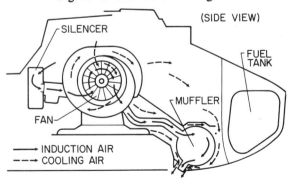

Fig. 11-16 Induction air mixes with fuel and moves through the engine to cool the rotor. Cooling air is blown across the radiating fins on the housing.

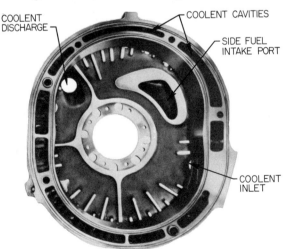

Fig. 11-17 Internal ribbing, side intake port, and coolant cavities of end housing (enclosing wall of end housing is machined away to show internal ribbing).

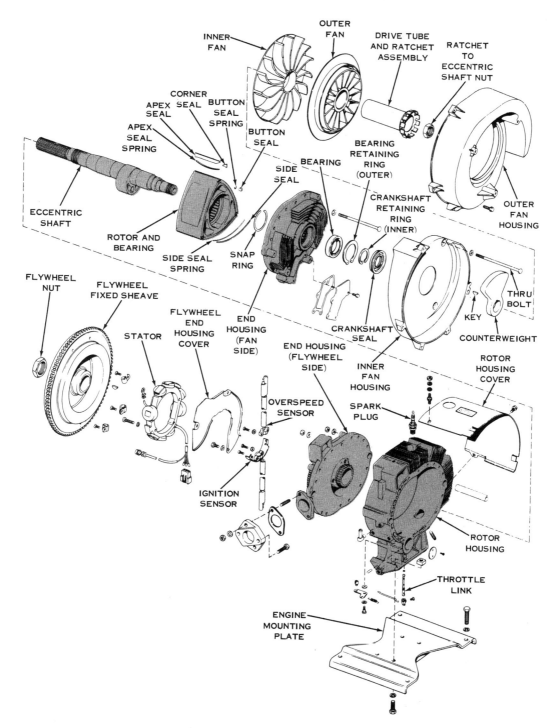

Fig. 11-18 Assembly drawing of the OMC rotary combustion engine (Basic parts of the engine are shaded.)

rotor. The rotor housing working surface is plated with a mixture of steel and bronze, while the apex seals are gray cast iron.

Lubrication in the OMC engine is accomplished by mixing oil with gasoline in an oil-to-gasoline ratio of 1 to 50. The OMC engine uses a tungsten carbide coating on the rotor housing, steel apex seals, and cast iron side rotor seals; the basic castings are aluminum.

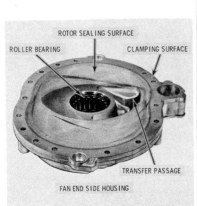
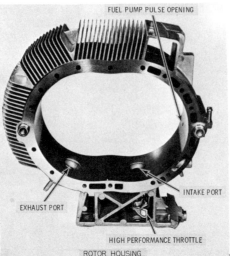
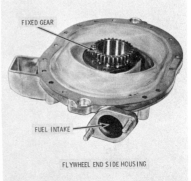

Fig. 11-19 Three main housings of the OMC RC engine showing the intake, transfer, and porting arrangements.

Conventional magneto ignitions and basic carburetors are used in the Fichtel and Sachs engines. The OMC engine employs a magneto-excited capacitor discharge ignition.

ROTATING COMBUSTION ENGINE PERFORMANCE

The performance of rotating combustion engines compares favorably to that of reciprocating engines. Performance characteristics such as brake horsepower, thermal efficiency, and mechanical efficiency, figure 11-20, indicate the adequacy of this engine. The torque curve for the RC engine is excellent with little drop at high speed. The flow of fuel mixture, demonstrated by the volumetric efficiency of the engine, is also excellent.

RC engines can use fuel with a lower octane rating than piston engines of the same compression ratio. The design of the combustion area and the flow of the fuel mixture lessens the possibility of preignition.

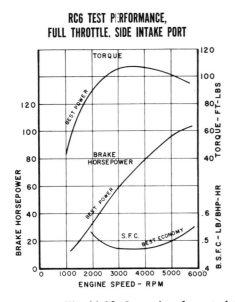
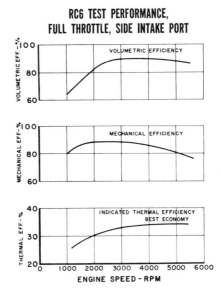

Fig. 11-20 Operating characteristics of a rotating combustion engine

Gasoline may be regular grade, either leaded or unleaded.

Emissions of unburned hydrocarbons, carbon monoxide (CO), and nitrogen oxides are present in the exhaust of RC engines, just as in reciprocating engines. Hydrocarbon and CO emissions are relatively high but can be lessened by having a well-tuned engine. Also, the smaller engine size permits space for emission control devices such as a thermal reactor. Nitrogen oxide emissions are somewhat lower than those in reciprocating engines, probably because maximum temperatures are less in an RC engine. Emission control devices help RC engines to meet U.S. Federal and European emission standards.

The horsepower range of the RC engine is flexible. The Curtiss-Wright corporation constructed one single-rotor engine with a 782 brake horsepower rating, as well as auto sized and small sized engines. The OS/Graupner Wankel engine powers model airplanes, figure 11-21, and produces 0.67 HP at 16,000 RPM. The horsepower of any basic engine can be increased by merely adding rotors in succession. A twin-rotor engine is equivalent to a six-cylinder reciprocating engine. The combination of three or four rotors and their enclosing chamber presents no construction problems, and an increase in the number of rotors offers a smoother operation.

The following points summarize the advantages of rotating combustion engines. (Also refer to figure 11-22.)

- Smooth operation with little vibration.
- High horsepower-to-weight ratio.
- Many different fuels can be used.
- Wider speed range than that of reciprocating engines.
- Excellent torque curve.
- Economical to manufacture.
- Can be produced in many different sizes.
- Fewer moving parts than reciprocating engines.

Fig. 11-21 The OS/Graupner Wankel engine for model airplanes

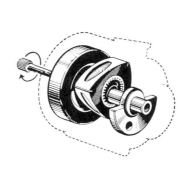

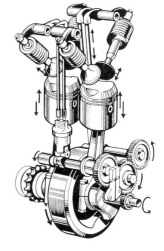

Fig. 11-22 Rotating combustion engines require fewer parts, are lighter in weight, and are more economical to manufacture than reciprocating engines.

REVIEW QUESTIONS

1. Explain the difference between rotating motion and reciprocating motion as these terms are applied to internal combustion engines.

2. What two significant problems slowed the development of rotating combustion engines?

3. Who is credited with perfecting the design for the most widely used rotating combustion engine?

4. What name is given to the geometric shape of the rotor housing?

5. What four functions must be accomplished during a complete cycle of the rotating combustion engine?

6. Does the rotating combustion engine require a unique or specially designed carburetor, ignition system, water pump, or oil pump? Explain.

7. List the key moving parts of the rotating combustion engine.

8. List several advantages of the rotating combustion engine.

CLASS DISCUSSION TOPICS

- Discuss the role of technology in developing and manufacturing the rotating combustion engine.
- Discuss the various problems that an emerging design encounters as it becomes an established and widely used product.
- Discuss the operating characteristics of the rotating combustion engine, such as torque, volumetric efficiency, mechanical efficiency, and horsepower.
- Discuss the operating principles of the rotating combustion engine.

LABORATORY EXPERIENCE 11-1
PREPARING THE RC ENGINE FOR SERVICE

Note: Many of the previous laboratory experiences apply to the RC (Wankel) engine. Using specifications for a particular RC engine model converts a laboratory experience from reciprocating to rotating combustion. For example, laboratory experience 4-1, Basic Parts of the Carburetor, satisfies both reciprocating and RC engines. If disassembly or overhaul of an RC engine is anticipated, it is recommended that all operations be done with the aid of an applicable service repair manual and expert supervision.

For this lab experience, refer to the following manuals: SACHS Wankel Engine KM 24 Manual 4016.2E Snowmobile Engine, and 1973 Evinrude and Johnson Snowmobile Service Manual 35 HP. Before starting any engine read and understand all instructional material. This helps ensure correct operation and adds to the service life of the engine.

Students are to perform the following steps on an RC engine. Instructors may want to revise or supplement certain steps to meet the needs of available engine models; the steps below are a general guide. These steps cover the engine itself, and there may be additional points to be covered on the machine the engine is powering.

1. Check the spark plug for tightness (12-15 ft.-lbs.). Be certain the spark plug gasket is in place.

2. Check the spark plug lead; is it secure to the spark plug?

3. Using a funnel with a fine-mesh screen, fill the gasoline tank with fuel mixture, 50 to 1 (gas to oil). Never fill the tank while the engine is running. Use regular gasoline, leaded or unleaded. OMC/RC engines use Evinrude and Johnson Rotary Combustion Snowmobile Lubricant or OMC Rotary Combustion Lubricant. Fichtel and Sachs engines use Sachs Rotary Piston Oil, BP Super Outboard Motor Oil, ESSO Lub HD 30, Mobilmix TT, Mobiloil TT, or Shell-Rotella SAE 30. When mixing, partially fill the mixing container with gasoline, add lubricating oil, shake vigorously to mix, add the remaining gasoline to obtain the proper ratio, and again shake vigorously to complete the mixing.

4. New engines require a break-in period. Operate a Sachs engine at about half throttle under medium RPM for the first 5 running hours. OMC engines should be operated at less than full throttle until at least one tankful of fuel has been used.

5. Understand engine safety and be familiar with all operating controls before attempting to start the engine and operate the machine.

REVIEW QUESTIONS

1. Explain what might happen if the lubricating oil is not mixed with the gasoline.

2. What is the correct ratio of gasoline to oil?

3. List safety considerations for operating the engine.

4. Explain the starting procedure for the engine.

5. Explain the operating controls for the machine, such as shifting gears, choke, lights, and safety features.

LABORATORY EXPERIENCE 11-2
MAINTENANCE OF THE RC ENGINE

For this lab experience, refer to the following manuals: SACHS Wankel Engine KM 24 Manual 4016.2E Snowmobile Engine, and 1973 Evinrude and Johnson Snowmobile Service Manual 35 HP. The best source of information on engine care is the manual or instruction booklet supplied with every new engine. This information tells the engine owner or operator about the requirements of that particular engine.

Students are to perform several maintenance activities on an RC engine. Instructors may want to revise or supplement certain steps to meet the requirements of available engine models.

1. Remove the spark plug and clean it. Carbon deposits must be carefully removed, preferably with an abrasive blast cleaning machine. For Sachs engines, clean the plugs after every 100 hours of operation.

2. Lubricate the recoil starter as required if rewinding trouble develops. For Sachs engines, use 0.5 cc of Anticorit oil at the grease fitting.

3. Steps 3 and 4 should be done with the engine stopped and cold. Gasoline spilled on a hot engine can explode. Remove the fuel strainer of the fuel pump and clean as required in solvent.

4. Clean the fuel strainer at the tank if one is installed.

5. Remove and clean the air cleaner as needed (for OMC engines, twice a season).

REVIEW QUESTIONS

1. Explain the difference between preventive maintenance and repair.

2. Describe the condition of the spark plug.

3. What type of spark plug is recommended for the engine?

4. What type of air cleaner is used on the engine?

LABORATORY EXPERIENCE 11-3
PREPARING THE RC ENGINE FOR STORAGE

For this lab experience, refer to the following manuals: SACHS Wankel Engine KM 24 Manual 4016.2E Snowmobile Engine, and 1973 Evinrude and Johnson Snowmobile Service Manual 35 HP.

> Note: Engines that are operated on a seasonal basis should have special care during their off-season storage. To ensure long, dependable engine life, follow the suggestions of the engine manufacturer.

Students are to perform the following steps in preparing an RC engine for storage. Instructors may want to revise or supplement the list to meet the needs of a particular engine or machine.

A. For SACHS engines:

1. Squirt 20 cc (1.22 cu. in.) of oil (Shell ENSIS 30) into the carburetor inlet with the choke and throttle open. Crank the engine 5 or 6 times to spread the oil on the bearings, rotor seals, and working surfaces of the end housings.

2. One of the following anticorrosive oils should be applied to the outside surfaces:
   Anticorit 5 or Messrs FUCHS D-6800
   Mobil MIL-L-644B
   Shell ENSIS Fluid 260
   ESSO Rust Ban 395

3. Drain the fuel system of the engine.

B. For OMC engines:

1. Drain the fuel tank, using a siphon hose.

2. Operate the engine to consume the fuel remaining in the carburetor.

3. Remove the spark plug.

4. Insert 1 teaspoon of OMC Rotary Combustion Oil into the engine through the spark plug hole. Move the eccentric shaft one revolution. Insert a teaspoon of oil to the second chamber and again turn the shaft one revolution. Repeat for the third chamber. Now crank the engine several times to distribute the oil. Replace the spark plug.

5. Apply OMC Accessory Engine Cleaner to the engine.

6. Clean or replace the fuel pump filter screen.

7. After removing the air cleaner element, clean or replace it.

8. Perform additional storage requirements by consulting the manufacturer's instruction book for the snowmobile.

REVIEW QUESTIONS

1. What can happen to gasoline during long periods of storage?

2. Can engine winterizing be accomplished by the owner? If it can, what information sources can be employed?

APPENDIX

RECIPROCATING ENGINE
TROUBLESHOOTING CHART

ENGINE FAILS TO START OR STARTS WITH DIFFICULTY

Cause	Remedy
No fuel in tank	Fill tank with clean, fresh fuel.
Shutoff valve closed	Open valve.
Obstructed fuel line	Clean fuel screen and line. If necessary, remove and clean carburetor.
Tank cap vent obstructed	Open vent in fuel tank cap.
Water in fuel	Drain tank. Clean carburetor and fuel lines. Dry spark plug points. Fill tank with clean, fresh fuel.
Engine overchoked	Close fuel shutoff and pull the starter until engine starts. Reopen fuel shutoff for normal fuel flow.
Improper carburetor adjustment	Adjust carburetor.
Loose or defective magneto wiring	Check magneto wiring for shorts or grounds; repair if necessary.
Faulty magneto	Check timing, point gap, and if necessary, overhaul magneto.
Spark plug fouled	Clean and regap spark plug.
Spark plug porcelain cracked	Replace spark plug.
Poor compression	Overhaul engine.

ENGINE KNOCKS

Cause	Remedy
Carbon in combustion chamber	Remove cylinder head or cylinder and clean carbon from head and piston.
Loose or worn connecting rod	Replace connecting rod.
Loose flywheel	Check flywheel key and keyway; replace parts if necessary. Tighten flywheel nut to proper torque.
Worn cylinder	Replace cylinder.
Improper magneto timing	Time magneto.

ENGINE MISSES UNDER LOAD

Cause	Remedy
Spark plug fouled	Clean and regap spark plug.
Spark plug porcelain cracked	Replace spark plug.
Improper spark plug gap	Regap spark plug.
Pitted magneto breaker points	Clean and dress breaker points. Replace badly pitted breaker points.
Magneto breaker arm sluggish	Clean and lubricate breaker point arm.
Faulty condenser (except on Tecumseh Magneto)	Check condenser on a tester; replace if defective.
Improper carburetor adjustment	Adjust carburetor.
Improper valve clearance (four-stroke cycle engines)	Adjust valve clearance.
Weak valve spring (four-stroke cycle engines)	Replace valve spring.
Reed fouled or sluggish (two-stroke cycle engines)	Clean or replace reed.
Crankcase seal leak (two-stroke cycle engines)	Replace worn crankcase seals.

ENGINE LACKS POWER

Cause	Remedy
Choke partly closed	Open choke.
Improper carburetor adjustment	Adjust carburetor.
Magneto improperly timed	Time magneto.
Worn piston or rings	Replace piston or rings.
Lack of lubrication (four-stroke cycle engine)	Fill crankcase to proper level.
Air cleaner fouled	Clean air cleaner.
Valves leaking (four-stroke cycle engine)	Grind valves.
Reed fouled or sluggish (two-stroke cycle engine)	Clean or replace reed.
Improper amount of oil in fuel mixture (two-stroke cycle engine)	Drain tank; fill with correct mixture.
Crankcase seals leaking (two-stroke cycle engine)	Replace worn crankcase seals.

ENGINE OVERHEATS

Cause	Remedy
Engine improperly timed	Time engine.
Carburetor improperly adjusted	Adjust carburetor.
Airflow obstructed	Remove any obstructions from air passages in shrouds.
Cooling fins clogged	Clean cooling fins.
Excessive load on engine	Check operation of associated equipment. Reduce excessive load.
Carbon in combustion chamber	Remove cylinder head or cylinder and clean carbon from head and piston.
Lack of lubrication	Fill crankcase to proper level.
Improper amount of oil in fuel mixture	Drain tank; fill with correct mixture.

ENGINE SURGES OR RUNS UNEVENLY

Cause	Remedy
Fuel tank cap vent hole clogged	Open vent hole.
Governor parts sticking or binding or bent	Clean, and if necessary, repair governor parts.
Carburetor throttle linkage, throttle shaft and/or butterfly binding or sticking	Clean, lubricate, or adjust linkage and deburr throttle shaft or butterfly.

ENGINE VIBRATES EXCESSIVELY

Cause	Remedy
Engine not securely mounted	Tighten loose mounting bolts.
Bent crankshaft	Replace crankshaft.
Associated equipment out of balance	Check associated equipment.

ROTATING COMBUSTION
ENGINE TROUBLESHOOTING CHART*

Troubleshooting to determine the cause of any operating problem may be broken down into the following steps:

a. Obtaining an accurate description of the trouble

b. Preliminary inspection

c. Use of Trouble Check Chart to analyze engine performance

An accurate description of the trouble is essential for troubleshooting. The owner's comments may provide valuable information which will serve as a clue to the cause of the problem.

Preliminary Inspection

1. Engine turns over freely

2. Spark at spark plug

3. Carburetor adjusted properly

4. Power port butterfly closed completely at idle

5. Air cleaner clean and properly installed

6. Air intake clean and not restricted

7. Cooling fan intake not restricted and fan air exit under muffler not restricted

8. Proper fuel and oil in tank

9. Static air leak check within limits

STARTING

1. Hard to start or won't start

    a. Low static air leak check

    b. Old fuel, water in fuel, or restricted fuel passages

    c. Choke not working properly

    d. Primer not working properly

    e. Carburetor low-speed needle not adjusted correctly

    f. Bad spark plug

    g. No spark

    h. Air cleaner or intake blocked or restricted

    i. Clogged fuel line or filter

    j. Engine flooded

    k. Faulty or missing gaskets on intake systems

2. Engine turns over extremely easily

    a. Low static air leak check

    b. Spark plug loose

    c. Cracked or broken engine castings

3. Engine starts, but stops immediately

    a. Fuel pump not working

    b. Carburetor low-speed system blocked or out of adjustment

    c. Fuel filter clogged

    d. Choke not working properly

4. Engine won't turn over

    a. External parts assembled wrong causing interference

    b. Rotor assembled upside down, gears not meshing

    c. "J" gap plug installed instead of surface gas type

    d. No gasket under surface gap plug

    e. Bearings seized

    f. Rotor cracked or broken

    g. Engine castings cracked or broken

5. Weak spark or no spark

    a. Faulty charge coil

    b. Faulty sensor coil

    c. Faulty power pack

    d. Grounded ignition switch or wire

    e. Flywheel not magnetized

    f. Faulty ignition coil or leads

    g. Ignition switch not working

STARTING—MANUAL STARTER

1. Manual starter pulls out, but starter does not engage flywheel

    a. Excess or incorrect grease on pawls or spring

    b. Pawls bent or burred

    c. Pawls frozen (water) in place

2. Starter rope does not return

    a. Recoil spring broken or binding

    b. Starter housing bent

    c. Loose or missing parts

3. Clattering manual starter

    a. Friction spring bent or burred

    b. Starter housing bent

    c. Excess or incorrect grease on pawls or spring

    d. Dry starter spindle

STARTING—ELECTRIC STARTER

1. Starter cranks too slowly

    a. Weak battery

    b. Loose or corroded connections or ground connection

    c. Faulty starter solenoid or solenoid wiring

    d. Worn armature brushes or spring

    e. Faulty field or armature (shorted or open windings)

2. Starter will not crank engine

    a. Weak battery

    b. Loose or corroded connections or ground

    c. Broken wire in harness or connector

    d. Faulty ignition key switch

    e. Faulty starter solenoid or solenoid wiring

    f. Moisture in starter motor

    g. Broken or worn brushes or broken brush spring

    h. Faulty field or armature (shorted or open windings)

    i. Starter does not engage with engine because drive gear is not free on helix.

3. Starter will not disengage flywheel ring gear

    a. Drive gear is not free on helix (debris must be removed)

    b. Lubricate helix

RUNNING—LOW SPEED

1. Low-speed miss

    a. Incorrect gas-lubricant ratio

    b. Incorrect idle adjustment

    c. Loose or broken ignition coil wires

    d. Spark plug terminal loose

    e. Weak coil

    f. Loose electrical connections

    g. Power port butterfly not closed completely

    h. Bad or missing gaskets around intake manifold

    i. Choke not operating correctly

    j. Low static air leak check

RUNNING—HIGH SPEED

1. High-speed miss

    a. Overspeed sensor improperly adjusted

    b. Water in fuel

    c. Weak spark

    d. Arcing around ignition coil or leads

    e. Bad spark plug

    f. Low static air leak check

    g. Carburetor inlet needle sticking

2. Poor acceleration, top RPM is low

    a. Incorrect gas-lubricant ratio

    b. Old fuel

    c. Fuel hose plugged or kinked

    d. Fuel filter restricted

    e. Bad fuel pump

    f. Pulse line to fuel pump restricted

    g. Loose or broken high-tension lead

    h. Weak coil

    i. Bad power pack

    j. Carburetor passageways restricted

    k. Power port butterfly not opening completely

    l. Overheating

    m. Low static air leak check

    n. Fuel tank vent restricted or blocked

RUNNING—HIGH AND LOW SPEED

1. Engine overheats

    a. Incorrect gas-lubricant ratio

    b. Improper engine assembly

    c. Cooling fins blocked by foreign material

    d. Cooling fan intake restricted

    e. Cooling air exit restricted

    f. Dirty air filter

    g. Air intake restricted

    h. Intake air box leaking, getting air from under shroud

2. Engine seizes (stops suddenly)

    a. No oil in gas

    b. Seized rotor or main bearing

    c. Broken rotor or stationary gear

    d. Cracked or broken engine castings

*Courtesy of Outboard Marine Corp.

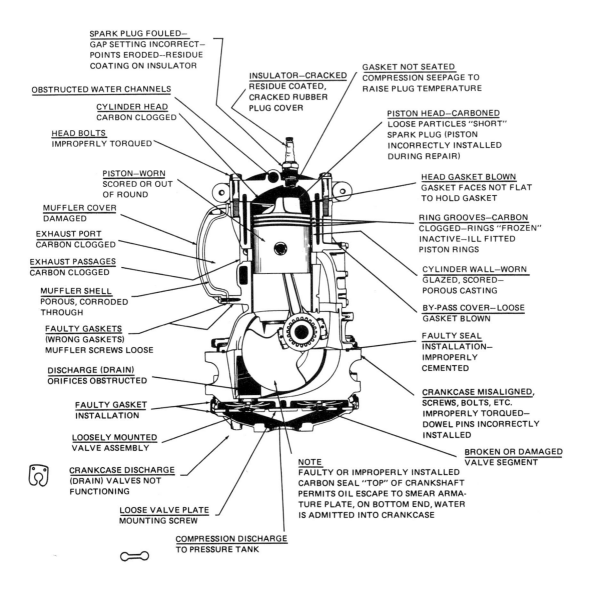

SPARK PLUG FOULED—
GAP SETTING INCORRECT—
POINTS ERODED—RESIDUE
COATING ON INSULATOR

INSULATOR—CRACKED
RESIDUE COATED,
CRACKED RUBBER
PLUG COVER

GASKET NOT SEATED
COMPRESSION SEEPAGE TO
RAISE PLUG TEMPERATURE

OBSTRUCTED WATER CHANNELS

CYLINDER HEAD
CARBON CLOGGED

HEAD BOLTS
IMPROPERLY TORQUED

PISTON HEAD—CARBONED
LOOSE PARTICLES "SHORT"
SPARK PLUG (PISTON
INCORRECTLY INSTALLED
DURING REPAIR)

PISTON—WORN
SCORED OR OUT
OF ROUND

HEAD GASKET BLOWN
GASKET FACES NOT FLAT
TO HOLD GASKET

MUFFLER COVER
DAMAGED

RING GROOVES—CARBON
CLOGGED—RINGS "FROZEN"
INACTIVE—ILL FITTED
PISTON RINGS

EXHAUST PORT
CARBON CLOGGED

EXHAUST PASSAGES
CARBON CLOGGED

CYLINDER WALL—WORN
GLAZED, SCORED—
POROUS CASTING

MUFFLER SHELL
POROUS, CORRODED
THROUGH

BY-PASS COVER—LOOSE
GASKET BLOWN

FAULTY GASKETS
(WRONG GASKETS)
MUFFLER SCREWS LOOSE

FAULTY SEAL
INSTALLATION—
IMPROPERLY
CEMENTED

DISCHARGE (DRAIN)
ORIFICES OBSTRUCTED

FAULTY GASKET
INSTALLATION

CRANKCASE MISALIGNED,
SCREWS, BOLTS, ETC.
IMPROPERLY TORQUED—
DOWEL PINS INCORRECTLY
INSTALLED

LOOSELY MOUNTED
VALVE ASSEMBLY

BROKEN OR DAMAGED
VALVE SEGMENT

CRANKCASE DISCHARGE
(DRAIN) VALVES NOT
FUNCTIONING

NOTE
FAULTY OR IMPROPERLY INSTALLED
CARBON SEAL "TOP" OF CRANKSHAFT
PERMITS OIL ESCAPE TO SMEAR ARMA-
TURE PLATE, ON BOTTOM END, WATER
IS ADMITTED INTO CRANKCASE

LOOSE VALVE PLATE
MOUNTING SCREW

COMPRESSION DISCHARGE
TO PRESSURE TANK

Fig. A-1 Power head diagnosis chart

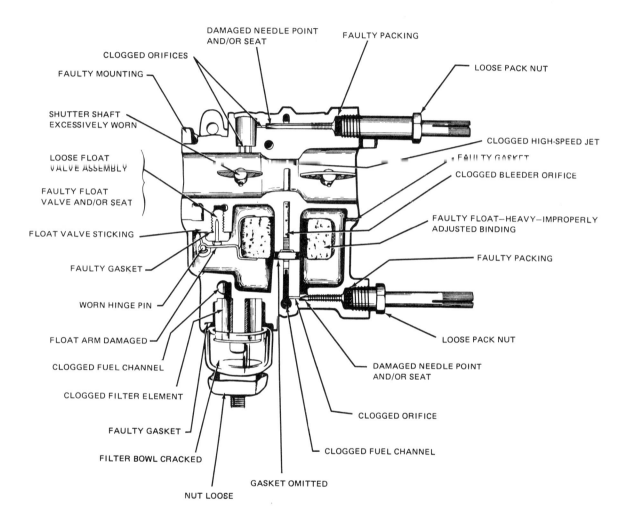

DAMAGED NEEDLE POINT AND/OR SEAT

FAULTY PACKING

CLOGGED ORIFICES

FAULTY MOUNTING

LOOSE PACK NUT

SHUTTER SHAFT EXCESSIVELY WORN

CLOGGED HIGH-SPEED JET

FAULTY GASKET

LOOSE FLOAT VALVE ASSEMBLY

CLOGGED BLEEDER ORIFICE

FAULTY FLOAT VALVE AND/OR SEAT

FAULTY FLOAT—HEAVY—IMPROPERLY ADJUSTED BINDING

FLOAT VALVE STICKING

FAULTY PACKING

FAULTY GASKET

WORN HINGE PIN

LOOSE PACK NUT

FLOAT ARM DAMAGED

DAMAGED NEEDLE POINT AND/OR SEAT

CLOGGED FUEL CHANNEL

CLOGGED FILTER ELEMENT

CLOGGED ORIFICE

FAULTY GASKET

FILTER BOWL CRACKED

CLOGGED FUEL CHANNEL

NUT LOOSE

GASKET OMITTED

Fig. A-2 Carburetion diagnosis chart

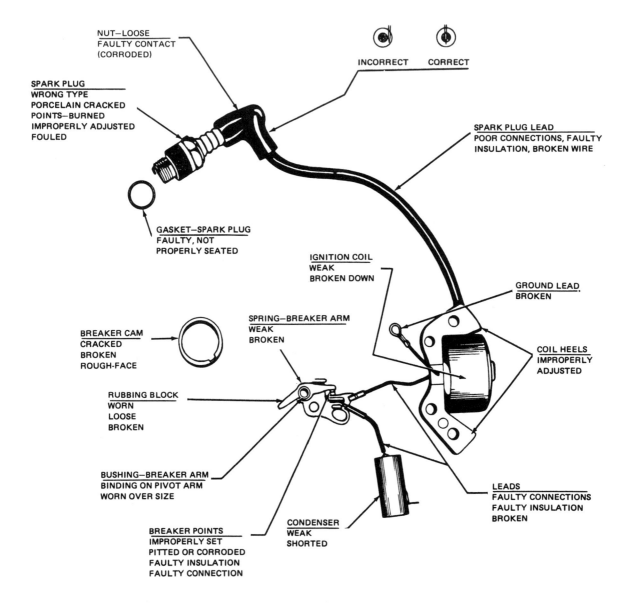

Fig. A-3 Magneto diagnosis chart

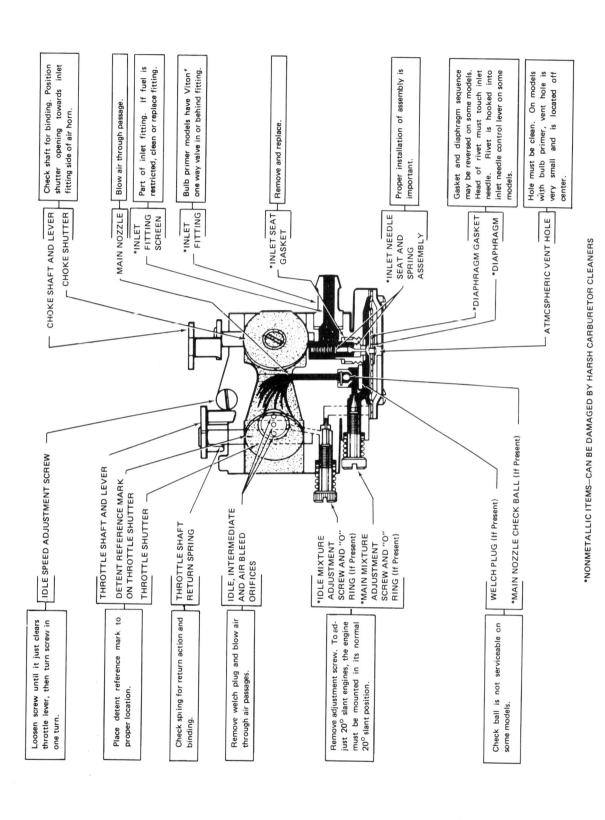

Fig. A-4 Service hints for diaphragm carburetors

CHOKE SHAFT AND LEVER — Check shaft for binding. Position shutter opening towards inlet fitting side of air horn.

CHOKE SHUTTER

MAIN NOZZLE — Blow air through passage.

*INLET FITTING SCREEN — Part of inlet fitting. If fuel is restricted, clean or replace fitting.

INLET FITTING — Bulb primer models have Viton one way valve in or behind fitting.

*INLET SEAT GASKET — Remove and replace.

*INLET NEEDLE SEAT AND SPRING ASSEMBLY — Proper installation of assembly is important.

*DIAPHRAGM GASKET — Gasket and diaphragm sequence may be reversed on some models. Head of rivet must touch inlet needle. Rivet is hooked into inlet needle control lever on some models.

*DIAPHRAGM

ATMCSPHERIC VENT HOLE — Hole must be clean. On models with bulb primer, vent hole is very small and is located off center.

IDLE SPEED ADJUSTMENT SCREW — Loosen screw until it just clears throttle lever, then turn screw in one turn.

THROTTLE SHAFT AND LEVER

DETENT REFERENCE MARK ON THROTTLE SHUTTER — Place detent reference mark to proper location.

THROTTLE SHUTTER

THROTTLE SHAFT RETURN SPRING — Check spring for return action and binding.

IDLE, INTERMEDIATE AND AIR BLEED ORIFICES — Remove welch plug and blow air through air passages.

*IDLE MIXTURE ADJUSTMENT SCREW AND "O" RING (If Present)

*MAIN MIXTURE ADJUSTMENT SCREW AND "O" RING (If Present) — Remove adjustment screw. To adjust 20° slant engines, the engine must be mounted in its normal 20° slant position.

WELCH PLUG (If Present)

*MAIN NOZZLE CHECK BALL (If Present) — Check ball is not serviceable on some models.

*NONMETALLIC ITEMS—CAN BE DAMAGED BY HARSH CARBURETOR CLEANERS

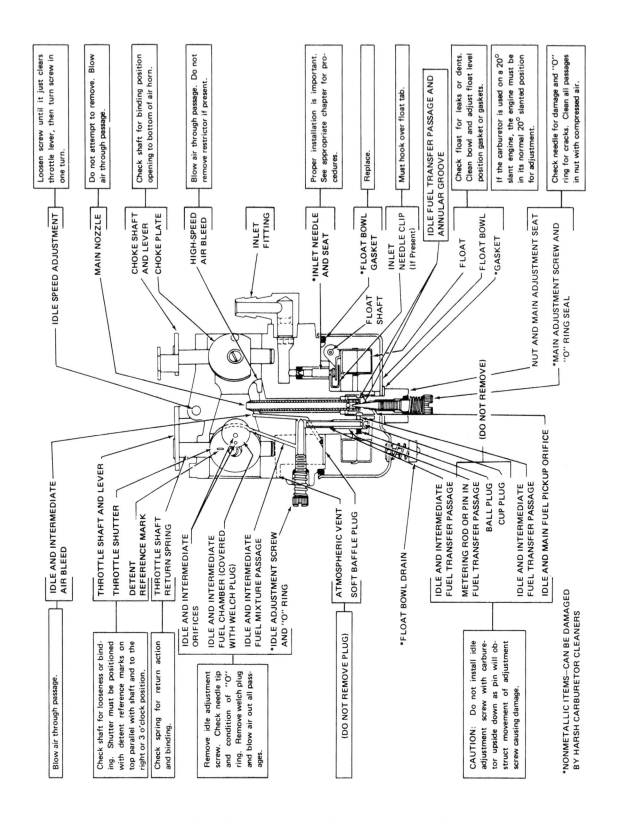

Fig. A-5 Service hints for float-feed carburetors

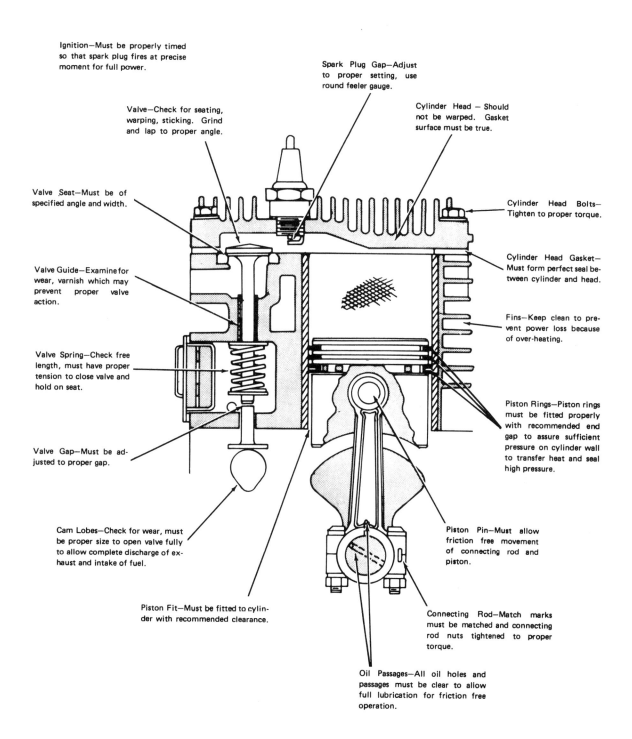

Ignition—Must be properly timed so that spark plug fires at precise moment for full power.

Spark Plug Gap—Adjust to proper setting, use round feeler gauge.

Valve—Check for seating, warping, sticking. Grind and lap to proper angle.

Cylinder Head — Should not be warped. Gasket surface must be true.

Valve Seat—Must be of specified angle and width.

Cylinder Head Bolts— Tighten to proper torque.

Valve Guide—Examine for wear, varnish which may prevent proper valve action.

Cylinder Head Gasket— Must form perfect seal between cylinder and head.

Fins—Keep clean to prevent power loss because of over-heating.

Valve Spring—Check free length, must have proper tension to close valve and hold on seat.

Piston Rings—Piston rings must be fitted properly with recommended end gap to assure sufficient pressure on cylinder wall to transfer heat and seal high pressure.

Valve Gap—Must be adjusted to proper gap.

Cam Lobes—Check for wear, must be proper size to open valve fully to allow complete discharge of exhaust and intake of fuel.

Piston Pin—Must allow friction free movement of connecting rod and piston.

Piston Fit—Must be fitted to cylinder with recommended clearance.

Connecting Rod—Match marks must be matched and connecting rod nuts tightened to proper torque.

Oil Passages—All oil holes and passages must be clear to allow full lubrication for friction free operation.

Fig. A-6 Points to check for engine power

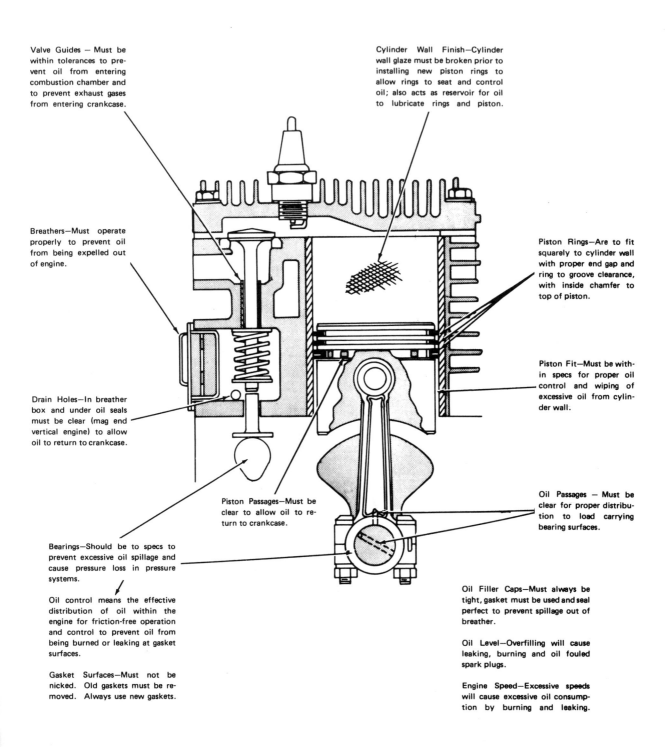

Valve Guides — Must be within tolerances to prevent oil from entering combustion chamber and to prevent exhaust gases from entering crankcase.

Cylinder Wall Finish—Cylinder wall glaze must be broken prior to installing new piston rings to allow rings to seat and control oil; also acts as reservoir for oil to lubricate rings and piston.

Breathers—Must operate properly to prevent oil from being expelled out of engine.

Piston Rings—Are to fit squarely to cylinder wall with proper end gap and ring to groove clearance, with inside chamfer to top of piston.

Drain Holes—In breather box and under oil seals must be clear (mag end vertical engine) to allow oil to return to crankcase.

Piston Fit—Must be within specs for proper oil control and wiping of excessive oil from cylinder wall.

Piston Passages—Must be clear to allow oil to return to crankcase.

Oil Passages — Must be clear for proper distribution to load carrying bearing surfaces.

Bearings—Should be to specs to prevent excessive oil spillage and cause pressure loss in pressure systems.

Oil control means the effective distribution of oil within the engine for friction-free operation and control to prevent oil from being burned or leaking at gasket surfaces.

Gasket Surfaces—Must not be nicked. Old gaskets must be removed. Always use new gaskets.

Oil Filler Caps—Must always be tight, gasket must be used and seal perfect to prevent spillage out of breather.

Oil Level—Overfilling will cause leaking, burning and oil fouled spark plugs.

Engine Speed—Excessive speeds will cause excessive oil consumption by burning and leaking.

Fig. A-7 Points to check for engine oil control

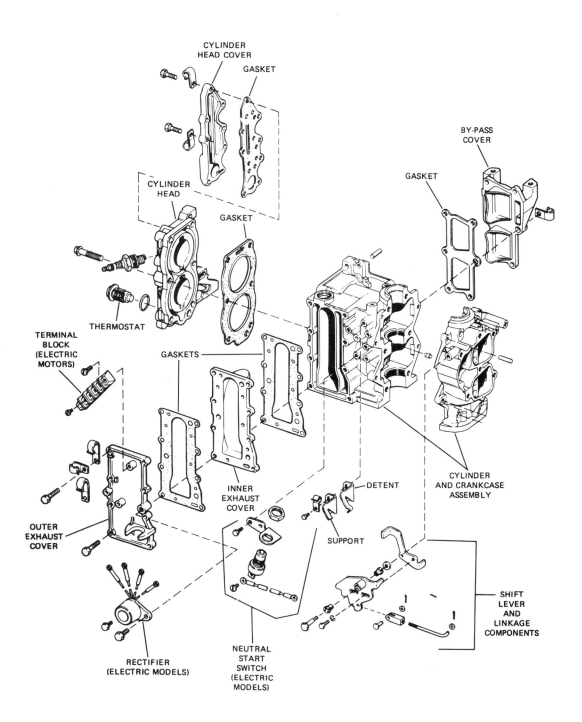

Fig. A-8 Power head—exploded view

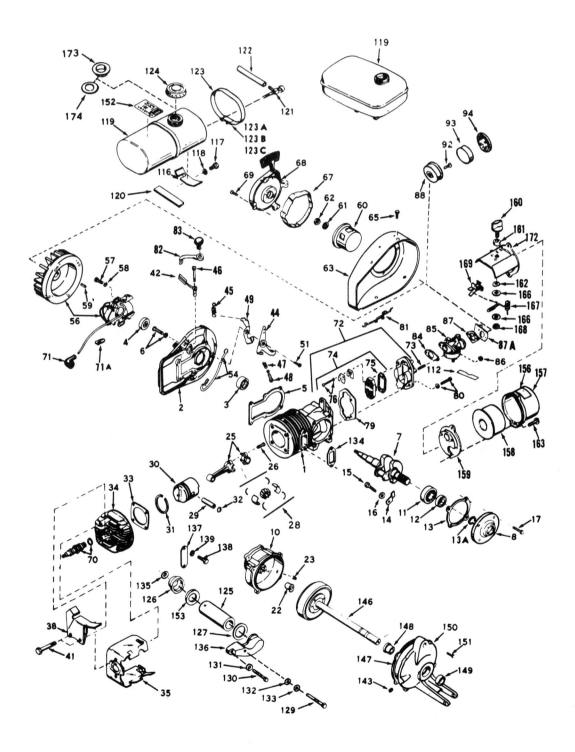

Fig. A-9 Horizontal engine, exploded view

1. Cylinder and Crankcase
2. Backplate with Bearing and Seal
3. Backplate Bearing
4. Backplate Oil Seal
5. Backplate Gasket
6. Screw and Lockwasher
7. Crankshaft
8. Crankcase Head
10. Crankcase Head
11. Bearing
12. Oil Seal
13. Crankcase Head Gasket
13A. Ring, Snap
14. Bearing Retainer Clip
15. Screw
16. Lockwasher
17. Screw
22. Gearcase Bushing
23. Plug
25. Connecting Rod Assembly
26. Screw
28. Bearing Assembly, 28 Rollers, 2 Liners, 4 Guides
29. Piston Pin
30. Piston
31. Piston Ring
32. Piston Pin Retaining Ring
33. Cylinder Head Gasket
34. Cylinder Head
35. Air Deflector
38. Fuel Tank Mounting Bracket
41. Screw
42. Governor Vane Assembly
44. Control Lever Assembly
45. Governor Vane Spring
46. Screw, Governor Vane
47. Spring, Control
48. Screw, Speed Control
49. Lever, Governor Spring
51. Screw, Control Lever Mounting
54. Stop Switch Wire

56. Magneto Assembly
57. Screw and Lockwasher
58. Washer
59. Flywheel Key
60. Rewind Starter Hub
61. Washer, Belleville
62. Nut
63. Fan Housing
65. Screw
67. Starter Screen
68. Rewind Starter Assembly
69. Screw and Lockwasher
70. Spark Plug
71. Spark Plug Cap
71A. Terminal, Spark Plug
72. Carburetor Adapter and Reed Plate Assembly
73. Stud
74. Reed Plate
75. Reed Plate Gasket
76. Screw, Reed Plate Mounting
79. Adapter Mounting Gasket
80. Screw and Lockwasher
81. Choke Link Assembly
82. Governor Link
83. Push Rivet
84. Carburetor Mounting Gasket
85. Carburetor Assembly
86. Nut
87. Plate Mounting Gasket
87A. Bracket, Throttle Wire
88. Air Filter Housing
92. Screw
93. Air Filter
94. Air Filter Cover
112. Decal, Choke, and Shutoff
116. Tank Mounting Bracket
117. Screw
118. Lockwasher (Washer)
119. Fuel Tank Assembly
120. Pad, Fuel Tank Mounting
121. Shutoff Valve Assembly

122. Fuel Line
123. Tank Mounting Strap
123A. Screw
123B. Lockwasher
123C. Nut
124. Fuel Tank Cap Assembly
125. Muffler
126. Muffler Cap
127. Muffler Mounting Gasket
128. Bolt
129. Screw
130. Screw
131. Lockwasher
132. Muffler Head Gasket
133. Washer
134. Exhaust Flange Gasket
135. Lockwasher
136. Muffler Head
137. Exhaust Flange Cover
138. Screw
139. Lockwasher
143. Plug
146. Gear and Shaft Assembly
147. Gear Reducer Cover Gasket
148. Gear Reducer Cover Bearing
149. Gear Reducer Cover Seal
150. Gear Reducer Cover
151. Screw
152. Decal, Name, Mix
153. Muffler Gasket
156. Decal, Air Cleaner
157. Air Filter Body
158. Air Filter
159. Bracket
160. Knob
161. Spacer
162. Washer
163. Screw
166. Washer
167. Spring
168. Nut
169. Stop Switch Assembly
172. Bracket Assembly
173. Baffle Assembly
174. Gasket, Fuel Tank Cap

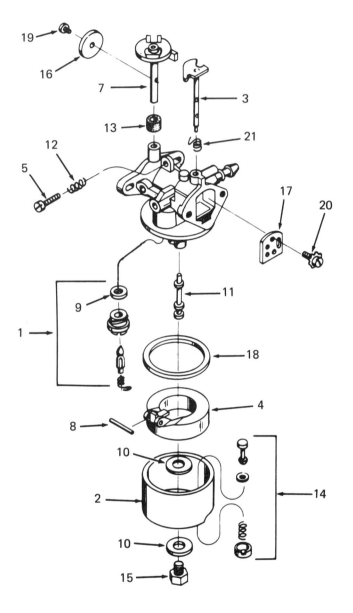

1. Matched, Float Valve Seat, Spring and Gasket Assembly
2. Bowl Assembly—Float Bowl
3. Shaft Assembly—Choke
4. Float Assembly
5. Screw—Throttle Adjustment
7. Shaft Assembly—Throttle
8. Shaft—Float
9. Gasket—Float Valve Seat
10. Gasket—Nut to Bowl
11. Main Metering Nozzle
12. Spring Throttle Adjustment
13. Seal—Throttle Shaft
14. Bowl Drain Assembly
15. Retainer Screw
16. Throttle Plate
17. Choke Plate
18. Gasket—Bowl to Body
19. Screw
20. Screw
21. Spring—Choke Return

Fig. A-10 Fixed-jet carburetor, exploded view

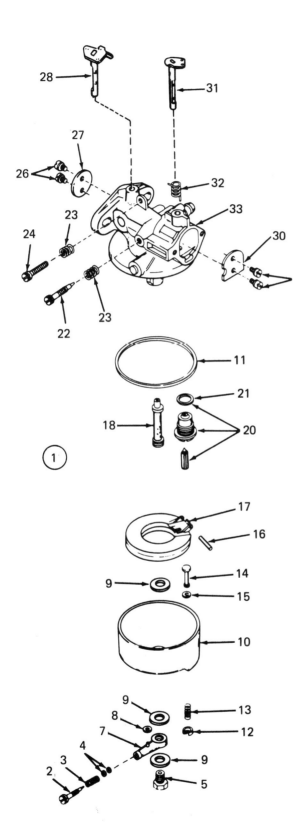

1. Carburetor Assembly
2. High-Speed Needle
3. Spring
4. "O" Rings
5. Retainer—Bowl
7. High-Speed Needle Housing
8. Rubber Gasket
9. Gasket—Bowl Nut to Bowl
10. Bowl Assembly—Float Bowl
11. Gasket—Body to Bowl
12. Retainer Screw
13. Spring—Drain Bowl
14. Stem Assembly—Drain Bowl
15. Rubber Gasket
16. Shaft—Float
17. Float Assembly
18. Main Metering Nozzle
20. Matched Float Valve, Seat, Spring and
    Gasket Assembly
21. Gasket—Seal Valve Float
22. Needle—Idle
23. Spring—Throttle Adjustment Screw
24. Screw—Idle Speed
26. Screw—Throttle Plate Mounting
27. Throttle Plate
28. Shaft Assembly—Throttle
29. Screw—Choke Plate Mounting
30. Choke Plate
31. Shaft Assembly—Choke
32. Choke Return Spring
33. Carburetor Body

Fig. A-11 Adjustable carburetor, type 1, exploded view

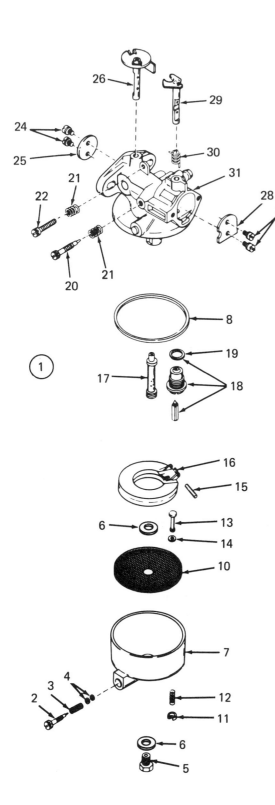

1. Carburetor Assembly
2. High-Speed Needle
3. Spring
4. "O" Rings
5. Retainer Screw
6. Gasket—Bowl Nut to Bowl
7. Bowl Assembly—Float Bowl
8. Gasket—Body to Bowl
10. Screen
11. Retainer Screw
12. Spring—Drain Bowl
13. Stem Assembly—Drain Bowl
14. Rubber Gasket
15. Shaft—Float
16. Float Assembly
17. Main Metering Nozzle
18. Matched Float Valve Seat, Spring and
    Gasket Assembly
19. Gasket—Seal Valve Float
20. Needle, Idle
21. Spring
22. Screw, Throttle Adjustment
24. Screw—Throttle Plate Mounting
25. Throttle Plate
26. Shaft Assembly Throttle
27. Screw—Choke Plate Mounting
28. Choke Plate
29. Shaft Assembly—Choke
30. Spring Choke Return
31. Carburetor Body

Fig. A-12 Adjustable carburetor, type 2, exploded view

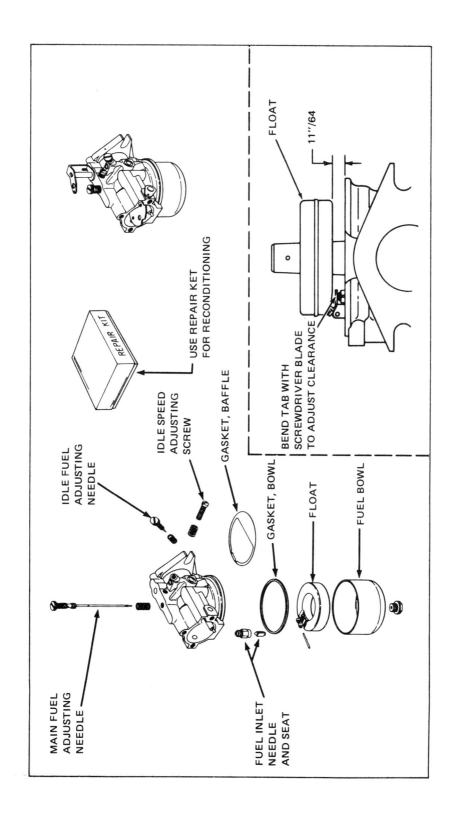

Fig. A-13 Side-draft carburetor

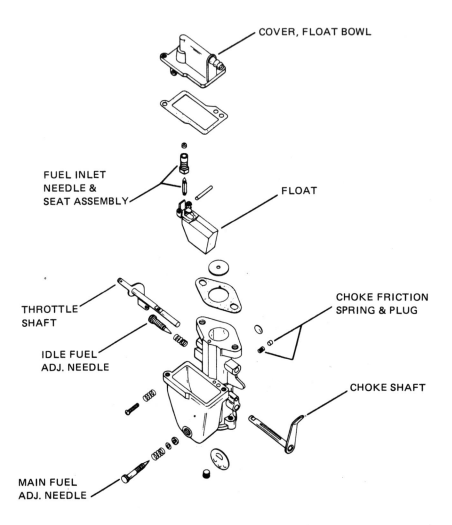

COVER, FLOAT BOWL

FUEL INLET
NEEDLE &
SEAT ASSEMBLY

FLOAT

THROTTLE
SHAFT

CHOKE FRICTION
SPRING & PLUG

IDLE FUEL
ADJ. NEEDLE

CHOKE SHAFT

MAIN FUEL
ADJ. NEEDLE

Fig. A-14 Up-draft carburetor

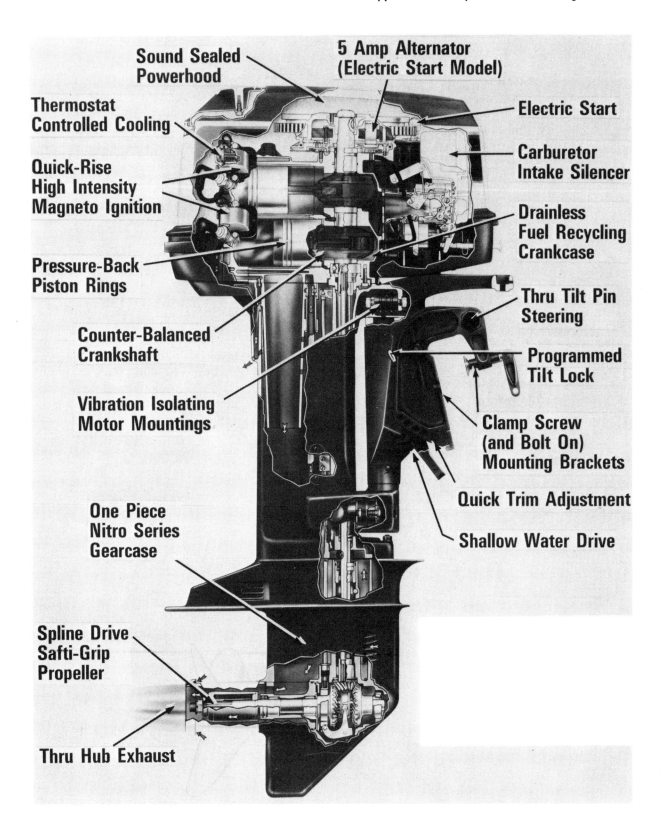

Sound Sealed Powerhood

5 Amp Alternator (Electric Start Model)

Electric Start

Thermostat Controlled Cooling

Carburetor Intake Silencer

Quick-Rise High Intensity Magneto Ignition

Drainless Fuel Recycling Crankcase

Pressure-Back Piston Rings

Thru Tilt Pin Steering

Counter-Balanced Crankshaft

Programmed Tilt Lock

Vibration Isolating Motor Mountings

Clamp Screw (and Bolt On) Mounting Brackets

Quick Trim Adjustment

One Piece Nitro Series Gearcase

Shallow Water Drive

Spline Drive Safti-Grip Propeller

Thru Hub Exhaust

Fig. A-15 Two-cylinder outboard engine

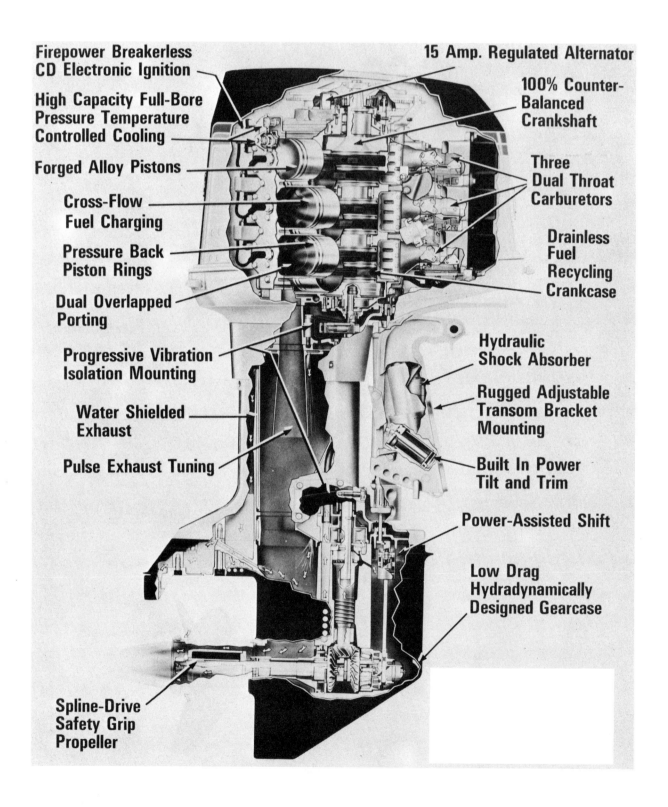

Firepower Breakerless CD Electronic Ignition

High Capacity Full-Bore Pressure Temperature Controlled Cooling

Forged Alloy Pistons

Cross-Flow Fuel Charging

Pressure Back Piston Rings

Dual Overlapped Porting

Progressive Vibration Isolation Mounting

Water Shielded Exhaust

Pulse Exhaust Tuning

Spline-Drive Safety Grip Propeller

15 Amp. Regulated Alternator

100% Counter-Balanced Crankshaft

Three Dual Throat Carburetors

Drainless Fuel Recycling Crankcase

Hydraulic Shock Absorber

Rugged Adjustable Transom Bracket Mounting

Built In Power Tilt and Trim

Power-Assisted Shift

Low Drag Hydradynamically Designed Gearcase

Fig. A-16 Six-cylinder outboard engine

ACKNOWLEDGMENTS

Contributions by Delmar Staff

Publications Director — Alan N. Knofla
Source Editor — Mark Huth
Associate Editor — Judith Barrow
Photography Editor — Sherry Patnode
Copy Editor — Noel Mick
Editorial Assistant — Virginia Styczynski

Manufacturers and government agencies have supplied information in the form of instruction books, books on basic principles, pamphlets, booklets, and photographs. Without these sources of firsthand information, the research and writing of this book would have been difficult, if not impossible.

The author wishes to acknowledge the following companies and organizations which have supplied material and information: Automotive Electric Association, Bolens Products Division Food Machinery and Chemical Corporation, Buick Motors Division General Motors Corporation, Cedar Rapids Engineering Company, Chrysler Corporation, Cushman Motor Works, Inc., Delco Remy Division General Motors, E. Edlemann and Company, Forster Brothers, Gravely Tractors, Inc., Holley Carburetor Company, Internal Combustion Engine Institute, Lawn Mower Institute, Inc., Marvel-Schebler Products Division Borg Warner Corporation, Mustang Motor Products Corporation, National Safety Council, Northrop Corporation, Oliver Outboard Motors, Outboard Motor Manufacturers' Association, Propulsion Engine Corporation, Quick Manufacturing, Inc., Thompson Products, Toro Manufacturing Corporation, Waukesha Motor Company, West Bend Aluminum Company, Westinghouse Electric Corporation, Whizzer Industries, Inc., Wright Saw Division Thomas Industries, Inc.

Particular thanks is due to those companies which have supplied many of the illustrations that are found in the text. Specific contributions are listed below.

AC Spark Plug Division, General Motors Corporation, Flint, Michigan — figures 7-29 and 7-30.
American Oil Company, Chicago Illinois — figure 4-27
Audi NSU Auto Union, 7107 Neckarsulm, German Federal Republic — figures 11-1, 11-2, 11-3, 11-4, 11-6, 11-14, 11-21, 11-22
Briggs and Stratton, Milwaukee, Wisconsin — figures 3-5, 3-19, 3-20, 3-28, 3-29, 3-37, 3-38, 4-1, 4-4, 4-6, 4-13, 4-14, 4-22, 4-23, 5-9, 5-15, 7-9, 7-20, 8-10, 9-17, 9-20, 9-23, 9-24, 9-29, 9-32, 9-33
Champion Spark Plug, Toledo, Ohio — figures 4-31, 4-32, 4-33, 7-28, 8-2, 8-3, 8-4, 8-5, 8-6

Clinton Engines, Clinton, Michigan — figures 3-6, 3-33, 5-4, 9-5, 9-8, 9-9, 9-10, 9-11, 9-12, 9-14, 9-16, 9-30, 9-35

Curtiss-Wright Corporation, Wood Ridge, New Jersey — figures 11-7, 11-8, 11-9, 11-11, 11-13, 11-17, 11-20

Ethyl Corporation, Ferndale, Michigan — figure 4-29

Evinrude, Milwaukee, Wisconsin — figures 6-10, A-15, A-16

Fairbanks-Morse, Chicago, Illinois — figures 7-8, 7-10, 7-11, 7-13, 7-14, 7-21, 7-23

General Motors, Detroit, Michigan — figures 2-4, 2-5, 2-6

Jacobsen Manufacturing Co., Racine, Wisconsin — figures 3-1, 8-13

Johnson Motors, Waukegan, Illinois — figures 3-2, 3-12, 3-13, 4-7, 4-8, 4-18, 4-30, 5-17, 6-7, 6-8, 6-9, 7-12, 7-22, A-1, A-2, A-3

Joseph Tardi Associates, Troy, New York — figures 2-3, 2-8, 2-9, 3-7, 3-14, 3-15, 3-22, 4-2, 4-3, 4-5, 4-10, 5-16, 8-8, 8-9, 8-14, 8-16, 9-13, 9-15, 9-22, 9-25, 9-26, 9-27

Kiekhafer Corporation, Fond du Lac, Wisconsin — figures 9-20, 9-28, 9-34, 9-36, 9-37, 9-38, 9-39

Kohler Company, Kohler, Wisconsin — figures 3-16, 3-17, 3-36, 6-3, 7-25

Lauson Power Products Department, Division of Tecumseh Products Company, Grafton, Wisconsin — figures 4-36, 4-37, 5-8, 5-12, 5-13

Lawn-Boy Company, Galesburg, Illinois — figures 3-3, 6-2, 9-7

Lincoln-Mercury Division, Ford Motor Company, Dearborn, Michigan — figures 3-4, 6-5, 6-6, 6-8, 8-7

McCulloch Company, Los Angeles, California — figures 3-35, 4-15, 4-16, 4-17, 4-20, 5-14, 7-18, 9-1

Mobil Oil Corporation, New York, New York — figures 11-5, 11-10, 11-12, 11-15

Outboard Marine Corporation, Waukegan, Illinois — figures 11-16, 11-18, 11-19

OMC-Lincoln, Lincoln, Nebraska — figures 3-31, 4-19, 5-5

Perfect Circle Company, Hagerstown, Indiana — figures 3-8, 3-9, 3-10, 9-31

Standard Oil Company of New Jersey, New York, New York — figures 3-23, 3-24, 3-26, 4-30, 5-7, 6-1

Tecumseh Products Company, Grafton, Wisconsin — figures 3-11, 3-27, 3-32, 3-34, 4-12, 4-26, 4-33, 5-6, 7-24, 7-26, 7-27, 8-11, 8-12, 8-15, 9-2, 9-3, 9-4, 9-18, 9-19, 9-21

Wico Electric Company, West Springfield, Massachusetts — figures 7-16, 7-17, 7-19

Wisconsin Motor Corporation, Milwaukee, Wisconsin — figures 4-35, 5-10, 5-11, 6-4

Zenith Carburetor Division, Bendix Corporation, Detroit, Michigan — figures 4-24, 4-25

INDEX